Opportunities and Uses
of the Ocean

David A. Ross

Opportunities and Uses of the Ocean

Springer-Verlag

New York   Heidelberg   Berlin

David A. Ross
Woods Hole Oceanographic Institution
Woods Hole, Massachusetts 02543
USA

With 144 figures.

Cover photograph courtesy of Dr. Robert Ballard, Woods Hole Oceanographic Institution. The picture was taken in the Galapagos Rift, a zone of active seafloor spreading, at a water depth of 2525 meters. The width of the crack is about 1 meter.

Library of Congress Cataloging in Publication Data

Ross, David A   1936-
    Opportunities and uses of the ocean.

    Bibliography: p.
    Includes index.
    1. Marine resources. I.   Title.
GC1015.2.R67   333.9′1   79-12694

Published 1980 by Springer-Verlag New York Inc.

Printed in the United States of America

9 8 7 6 5 4 3 2 1

ISBN 0-387-90448-4   Springer-Verlag   New York
ISBN 3-540-90448-4   Springer-Verlag   Berlin   Heidelberg

His Sea

Who hath desired the Sea?—the sight of salt water unbounded—
The heave and the halt and the hurl and the crash of the comber wind-hounded?
The sleek-barrelled swell before storm, grey, foamless, enormous, and growing—
Stark calm on the lap of the Line or the crazy-eyed hurricane blowing—
His Sea in no showing the same—his Sea and the same 'neath each showing:
        His Sea as she slackens or thrills?
So and no otherwise—so and no otherwise—hillmen desire their Hills!

Who hath desired the Sea?—the immense and contemptuous surges?
The shudder, the stumble, the swerve, as the star-stabbing bowsprit emerges?
The orderly clouds of the Trades, the ridged, roaring sapphire thereunder—
Underalded cliff-haunting flaws and the headsail's low-volleying thunder—
His Sea in no wonder the same—his Sea and the same through each wonder:
        His Sea as she rages or stills?
So and no otherwise—so and no otherwise—hillmen desire their Hills.

Who hath desired the Sea? Her menaces swift as her mercies?
The in-rolling walls of the fog and the silver-winged breeze that disperses?
The unstable mined berg going South and the calvings and groans that declare it—
White water half-guessed overside and the moon breaking timely to bare it;
His Sea as his fathers have dared—his Sea as his children shall dare it:
        His Sea as she serves him or kills?
So and no otherwise—so and no otherwise—hillmen desire their Hills.

Who hath desired the Sea? Her excellent loneliness rather
Than forecourts of kings, and her outermost pits than the streets where men gather
Inland, among dust, under trees—inland where the slayer may slay him—
Inland, out of reach of her arms, and the bosom whereon he must lay him—
His Sea from the first that betrayed—at the last that shall never betray him:
        His Sea that his being fulfils?
So and no otherwise—so and no otherwise—hillmen desire their Hills.

Rudyard Kipling

Preface

The oceans cover about 72 percent of our planet (which is named for the remaining 28 percent). These oceans have fascinated and challenged the human race for centuries. In the past, the ocean had been used first as a source of food and later as a means of transportation. However, the oceans have recently become very important—they may offer a solution to many of our modern problems. For example, refuse from land is to be dumped into the ocean never to be seen again; fish and other biological resources are to be caught and used to meet the protein deficiency of the world; oil and gas from the continental shelf and perhaps deeper areas will eventually solve our energy problems. None of these examples is completely possible, and the at-source of food and later as a means of transportation. However, the oceans social, and ecological problems in the marine environment. Countries are already planning how the oceans can be divided up, so that they will get their "fair share". Economists, politicians, and others are producing almost daily, optimistic or pessimistic views (depending upon your own viewpoint) about the ocean and its resources. Equally loud reports come from environmentalists, conservationists, government sources, and oil companies concerning the pollution and potential destruction of the ocean.

Where is the truth—what are the real problems and opportunities associated with the ocean? This book is written in an attempt to shed some light on how we now use the ocean, what are its opportunities and what we can do in the future. What I am trying to do is present some basic information about the ocean in a form that will be understandable to people who are not oceanographers, but who are interested and concerned about the marine environment.

The book has two obvious shortcomings. First, it is impossible to cover all aspects of all problems, but I do try to cover the important ones. In this regard I have probably overemphasized examples from the United States, but these are the ones I am most familiar with. Second, as a practicing oceanographer and occasionally involved in the area of marine policy and ocean management, I feel that I must take a position on some particular problems and opportunities. I don't apologize for this, although I try to indicate my own preferences or prejudices when I present them.

I have several people I want to thank. These include my colleagues who, although marine scientists at heart, often became interested in policy questions and in doing so have stimulated me to likewise think about these questions. These include people like Paul Fye, Ken Emery, Bob Morse, Bob Frosch, John Teal, Richard Haedrich, Charlie Hollister, John Knauss and Warren Wooster. Paul Fye encouraged me to get more involved in the Marine Policy and Ocean Management Program at the Woods Hole Oceanographic Institution, and through this program I made contacts with social scientists who were interested in the sea. I learned a lot from them and would especially like to acknowledge George Cadwalader, Susan Peterson, Leah Smith, Ed Miles, Tom Lechine and many other interesting people who participated in this program.

Several friends either read portions of the text, supplied me with pictures or both and these include Wayne Decker, Ken Emery, Bob Ballard, Jim Heirtzler, Robert Dinsmore, Phyllis Laking, Vicky Cullen, Jeffrey Ellis, John Ryther, Murray Felsher and Susumo Honjo. Leah Smith read the entire text and provided many valuable comments. Kathy Vermersch drafted some of the figures and Mavise Crocker and Ellen Gately assisted in the difficult job of keeping everything in order. My wife Edith, as she has done before, did many things that helped keep writing this book to a reasonable task.

Woods Hole,   David A. Ross
Massachusetts
October, 1979

Contents

Chapter 4

Marine Shipping   73

Chapter 5

The Resources of the Ocean   103

Chapter 6

Marine Pollution   194

Chapter 7

Military Uses of the Ocean   231

Chapter 1

The Opportunity of the Ocean

The future use and opportunities of the ocean are limited only by human imagination and technology. Whether the ocean will be used with wisdom and proper concern for its environment is only a hope, for past exploitation of the ocean shows a poor record. It has been rare that a local or state government has used the ocean, coastal zone, or rivers with a concern for another group or for another use of these resources. Our increased understanding of the marine environment clearly shows that this lack of concern has generally resulted in more overall harm than benefit.

Civilization has now developed to the point where almost any new technology, or any increased use of the environment, will have subsidiary effects—some of them unknown. The challenge is to optimize the good while either eliminating the bad or keeping it to a minimum. To do this for the marine environment requires a solid background of scientific knowledge and experience and the ability to perform experiments, to predict future effects, and to understand, as best as possible, the ramifications throughout the marine environment of whatever action is taken. Generally, such data and understanding are not presently available for the ocean. I should emphasize here that the marine environment is not a series of isolated phenomena, such as fish in one spot, plankton in another, and pollutants in a third; instead it is a complex and interacting system that is almost impossible to duplicate in a laboratory. What, then, is to be done? Shall those who want to use the coastal zone or other parts of the marine environment—the developers, fisherman, politicians, lawyers, swimmers, oilmen, and conservationists—wait until marine scientists get such data and understanding? The answer is obvious. The marine environment will continue to be used at an ever-increasing rate.

An important point in this discussion is that the environment is meant to be used, and that some risk is always involved. We breathe in oxygen from the atmosphere, convert some of it into carbon dioxide, and exhale it back into the atmosphere. This simple phenomenon would be catastrophic were it not that plants reconvert the carbon dioxide into oxygen and maintain the atmospheric balance. Risks are also present in the ocean but are usually more subtle. The ocean, like the atmosphere, is meant to be used; fish can be harvested from it, its minerals can be mined, and it can absorb some runoff and pollution from land. A risk-free condition is as impractical to achieve for the use of the ocean as it is for any environment. The ocean, for example, has an impressive capacity for absorbing material discharged into it from land. The rivers of the United States carry over 1.3 million tons of sediment (2,600,000,000 lb) every day into the ocean, usually without any major deleterious effects. This, in weight, is equivalent to dumping about 650,000 cars per day into the sea. The difference in the effect of putting cars or sediment into the near-shore zone of the ocean is not hard to visualize; however, surprisingly there can be situations where it might be better to have the cars than the sediment. The answers are not always obvious.

Some suggestions for the use of the oceans, such as underwater habitats to be used once the living space on land is gone, are just dreams with little chance of implementation. In this instance it would take several times the number of bottom occupants on the surface to maintain the habitat. Bottom habitats, however, can be a reality for the marine mining and petroleum industry where they could be used to monitor and control these operations. Other uses, such as the oceans being important energy and food sources or peaceful arenas for international exploration and exploitation, are viable possibilities.

The expenditure of money on items related to the use of the marine environment is immense. For example, about $125 billion is spent, worldwide, each year just for items related to outfitting, maintenance, overhaul, and replacement of ships and marine facilities, and for technical, engineering, and research activities. The present and future economic potentials of ocean resources are likewise impressive. Estimates of the present yearly value of marine resources (such as oil, gas, and fishery products) are over $60 billion and may double by the year 2000.

An indication of the importance of the ocean to a country such as the United States is that over 80% of its population lives within 150 km of the sea or the Great Lakes. This huge concentration of people presents numerous problems for the coastal zone, which unfortunately is one of the more fragile parts of the marine environment. Many coastal cities introduce large amounts of pollutants and waste products into rivers and coastal areas, which are also the breeding grounds or source areas of much of our important biologic resources, such as clams, oysters, and many fish.

The offshore areas adjacent to some large cities are often used as dumping grounds for industrial waste or human sewage. In most, if not all, situations there is insufficient knowledge of how these contaminants affect the environment or how the material decays or is redistributed. Indeed, it is not really known even whether some forms of waste may actually be of value to the environment, although generally they are not.

Shipping and the building of large tankers have increased at a remarkable rate over the last decade and similar growth is possible again in future years. These large ships require deep-water ports and new concepts for handling cargo. Harbors generally are badly situated, being where the land and water meet, an area where tides, waves, storms, and the shallowness of the water can cause considerable damage. As ships increase in size, should not offshore harbors, which are without most of the above problems, be used for the handling of cargo as well as for providing other innovative opportunities for the use of the sea? These offshore harbors could serve multiple purposes, such as sites for nuclear power plants (plenty of water for cooling and locations away from large concentrations of people), aquaculture sites (using heated water from the nuclear power plant), and airports. Other possibilities include a base for offshore fishing, drilling sites for petroleum, disposal of sewage (which could be used in an aquaculture system), and the like.

Another interesting but potentially devisive situation for the ocean is the Law of the Sea Conferences, which started with preparatory meetings in the early 1970s and are still continuing. These meetings are held in an attempt to redefine the legal aspects of the ocean. In the past, territorial seas were defined by the range of a cannon, i.e., a 3-nautical mile territorial sea within which the adjacent state had complete sovereign control. In 1945 the United States proclaimed jurisdiction over the resources on the continental shelf out to a depth of 200 m off its coast. The 200-m depth (656 ft is generally found at a distance considerably greater than 3 naut. mi. off the United States coast. This declaration by the United States was soon followed by other countries making unilateral claims, for both sea-bottom resources and fishing, out to as much as 200 miles off their coasts.

In 1958 a United Nations Conference on the Law of the Sea resulted in coastal states gaining sovereignty over the sea bed out to 200-m depth (if they signed the agreement). A controversial clause in the agreement indicated that additional deeper claims could be made as sea-bed resources become exploitable. In more recent years there has begun a "race" among developing and developed countries to extend their sovereignty seaward in a hope to include within them the real and imagined resources of the sea floor. An alternate view is that the resources of the ocean are the "common heritage of mankind" and should be shared equally by all countries. One of the main questions facing the present Law of the Sea Conference is how wide or narrow the territorial sea should be and what part of the ocean should be available for the international common heritage. Coastal states tend to favor a wide territorial sea. Landlocked states tend to favor narrow territorial seas and a broad international zone. Maritime states will want to preserve the broad and traditional freedoms of the sea. The decisions reached at the Law of the Sea Conference, if any, will also have important effects on freedom of scientific research, rights of innocent passage of ships, fisheries, and pollution, among other things. The conference also offers the countries of the world the opportunity to work together for their common good, because once the ocean is divided it is unlikely that it easily can be undivided.

Another major use and opportunity of the ocean concerns its mineral and biologic resources. Much has been written about these, and estimates of their worth vary by orders of magnitude. Certain facts are obvious: Energy resources, such as oil and gas, are to be found in many parts of the ocean, but probably not in the very deep sea. Presently, oil and gas exploration is occurring on the continental shelves of over 80 countries and within the decade the oil production from the marine environment could double. The recent awareness of an energy crisis has resulted in an intensification of exploration for offshore oil and gas resources. Manganese modules, found mainly in the deeper parts of the ocean, hold promise of considerable mineral value but present challenging legal problems concerning their ownership and technologic problems concerning their recovery and refining.

Some marine biologic resources may be overexploited and need multinational organizations or other mechanisms for their management, whereas others may not be fully used. A new potential resource, krill (a small shrimplike organism), may have a potential catch equal to that of all of the other biologic resources. In some instances marine resources are situated off a developing country and, if exploited properly, may be of considerable importance in the economic growth of the country. Conversely, many resources of the sea floor, such as nickel and cobalt from manganese modules, are already being produced by developing countries from land deposits; these countries therefore would be financially hurt by the marine mining of these resources by developed countries.

The oceans offer considerable opportunities for conflict. This can occur over their resources, over control of vital navigational channels, because of concern about pollution, or because of as yet unthought of reasons. Recent advances in oceanography and marine technology have undoubtedly contributed to the potential for conflict. However, these advances may also offer the opportunity for international cooperation. Within the past years several international organizations have been developed to handle and expedite such cooperation in fields of marine research, pollution monitoring of the environment, and resources development and management. Sophisticated instruments, such as satellites, have been used to collect data from the ocean concerning meteorologic conditions, biologic productivity, pollution effects, and other uses. These data are available to all countries.

The preceding paragraphs have presented in a general sense some of the uses, problems, and opportunities of the ocean; in following chapters I discuss the science of oceanography and then specific ocean uses, opportunities, and problems.

Chapter 2

How the Ocean Works

Introduction

The oceans are a relatively simple phenomenon when one considers their broad-scale features. They can be extremely complex, however, when details concerning oceanic processes are required. This is partially because we have been studying the oceans only for the last 100 years or so, and with sophisticated equipment for perhaps only the last 10 years, whereas many processes in the ocean take place over time scales that are considerably longer.

My objective in this chapter is to introduce the reader to the general aspects of the ocean and to the science of oceanography. Specific details can be obtained from any of the general oceanography texts I have listed at the end of the chapter. Perhaps the best place to start is to define oceanography: *the application of all science to the phenomena of the ocean.* Therefore, oceanography is really not a true science by itself but is a combination of sciences, especially geology, chemistry, biology, and physics. Many of the processes and reactions within the ocean reflect this mixture. An example is the geologic sampling of the sediments on the ocean bottom underlying the equatorial Pacific. The material there is mainly the shells of various species of microscopic plants and animals that have lived in the surface waters and have survived the slow trip, after their death, through the more than 3 miles of water to the ocean bottom. The presence of these shells on the bottom is caused by unique physical oceanographic conditions that exist in the surface waters and that in turn have affected the chemistry of the water, creating biologic conditions favorable for growth.

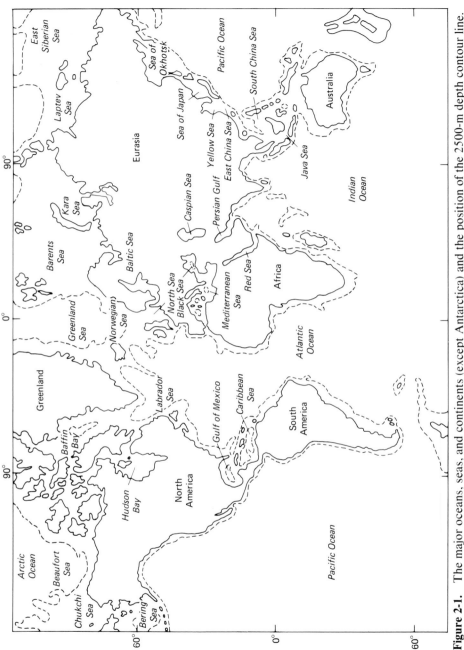

Figure 2-1.   The major oceans, seas, and continents (except Antarctica) and the position of the 2500-m depth contour line. Dashed line indicates the 2500-m depth contour, which is the approximate base of the continental slope.

General Characteristics of the Oceans

The surface of our planet is comprised of six continents—Africa, Eurasia, North America, South America, Australia, and Antarctica—which cover about 28% of the earth's surface; and four major oceans—the Pacific, Atlantic, Indian, and Arctic—plus numerous smaller seas, which comprise the remaining 72% (Fig. 2-1). Between the latitudes of about 55° and 65°S, essentially no land is present and the oceans encircle the entire globe. In general, the southern hemisphere can be considered as a water hemisphere, because most of the land is in the northern hemisphere.

The oceans are immense in their size, depth, and volume. Their average depth is 3729 m, or 12216 ft, or 2036 fathoms, or 2.3 miles (the average elevation of land is 840 m, or 2756 ft). Their area exceeds 362 000 000 km² or over 140 million square miles, and their volume is about 1 350 000 000 km³, or about 318 million cubic miles. The Pacific is the deepest and largest of the four major oceans (Table 2-1). Oceanographers, perhaps more than other scientists, use a mixture of measuring units (see Table 2-2).

The Origin of the Oceans

It is generally agreed that the earth became solidified about 4.5 billion years ago. At that time the oceans were not present. Actually, little is known about the early history of the ocean. Fossils of apparent marine organisms have been found and dated (by radioactive techniques) to be about 2 billion years old, suggesting that an ocean or, perhaps better said, a body of water existed at that time. Certainly the chemistry of that water body was different from that of the present ocean. Younger fossils, 600 million or so years in age, have been found that are similar to some present marine forms, suggesting that an ocean somewhat similar to the present one may have ex-

Table 2-1. Area, Volume, and Average Depth of the Oceans[a]

Oceans and Adjacent Seas	Area ($\times 10^6$ km²)	Volume ($\times 10^6$ km³)	Average Depth (m)
Arctic	12.257	13.702	1117
Indian	74.118	284.608	3840
Atlantic	94.314	337.210	3575
Pacific	181.344	714.410	3940
	362.033	1439.930	3729

[a]Data from Menard and Smith (1966).

Table 2-2. Some Commonly Used Oceanographic Units and Conversion Factors

To Convert	Into	Multiply by
Centimeters (cm)	Inches (in.)	0.3937
Meters (m)	Feet (ft)	3.28
Fathoms	Feet (ft)	6.0
Meters (m)	Fathoms	0.547
Kilometers (km)	Miles	0.621
km^2	Square miles	0.386
km^3	Cubic miles	0.24
Grams (g)	Ounces (avdp.)	0.035
Kilograms (kg)	Pounds (lb)	2.2
Degrees C	Degrees	($^\circ$C $\times$ 9/5) + 32
Liters	Gallons	0.2642
Grams/liter	Parts per million (ppm)	1 000

isted at that time. It should be emphasized that the shape and character of this early ocean basin did not resemble the present one (a point discussed in more detail in the section Origin of the Ocean Basin).

Two major questions exist concerning the origin of the ocean: (1) Where did the water come from? and (2) How did the ocean get its present chemical composition?

The best hypothesis of the origin of the water is that it has slowly been added to the ocean throughout geologic time, coming from volcanic activity, from heating and decomposition of volcanic rock (which contain about 5% water), and from hot springs. The elements in sea water are either cations (atoms or molecules that have a positive charge), such as calcium, sodium, or magnesium, or anions (atoms or molecules that have a negative charge), such as chloride or sulfide. It is thought that anions have been introduced into the ocean by volcanic activity, whereas cations come from the weathering of rocks on land or under the ocean.

The quantity of different elements in the ocean often bears little relationship to the rates at which they are introduced. This is because some elements, such as iron and aluminum, are very reactive in sea water and are quickly involved in chemical reactions, such as incorporation into the shell of an organism or precipitation to the bottom as part of the sediment, either of which removes them from the water. Other elements, such as sodium and magnesium, are essentially unreactive and remain in solution for long periods of time; in this manner the ocean has become relatively enriched in the unreactive elements.

All the reactions that influence the ocean's chemistry essentially take place at the interfaces or boundaries of the ocean. Basically, there are three such boundaries: (1) the water–atmosphere interface; (2) the water–sedi-

Table 2-3. The Distribution of the World's Water[a]

Environment	Water Volume (cubic miles)	Percentage of Total
Surface water		
Freshwater lakes	30 000	0.009
Saline lakes and inland seas	25 000	0.008
Rivers and streams	300	0.0001
Total	55 300	0.017
Subsurface water		
Soil moisture	16 000	0.005
Ground water	2 000 000	0.62
Total	2 016 000	0.625
Ice caps and glaciers	7 000 000	2.15
Atmosphere	3 100	0.001
Oceans	318 000 000	97.2
Totals (approx.)	326 000 000	100

[a]Data from Leopold and David (1966).

ment interface; and (3) the water–biosphere (or organism) interface. These are discussed further in later sections of this chapter.

It should be emphasized that most, but not all, of the water on the earth's surface is in the ocean (Table 2-3). Actually, about 97% of the water is in the ocean (and therefore salty), but if the water incorporated in the world's ice caps and glaciers were to melt, this percentage would increase to over 99. In addition, if the ice melted, sea level would rise 50 or 60 m and have a catastrophic effect on the coastal areas of the world.

Water does not necessarily remain forever in one of the categories shown in Table 2-3 but circulates between the atmosphere, oceans, land, and rivers. Water present as water vapor in the atmosphere probably only stays there for about 10 days before returning to the ocean or land as rain. This circulation pattern, whereby water moves from one environment to another, is called the hydrologic cycle (Fig. 2-2). This cycle is very important to such processes as the movement of pollutants. I will come back to some other important characteristics of sea water in a few pages.

The ocean, even though it has been around for 2 billion years or so, should not be considered as a static system. Actually it is dynamic in many aspects. There are changes that occur over the duration of thousands of years, such as glacial periods. There are also yearly climatic changes and monthly and daily changes related to weather and tides. Over geologic time, essentially all parts of the present continental areas have been covered by the ocean at least several times. As we shall see in a later section, the continents have also undergone considerable change and have moved around.

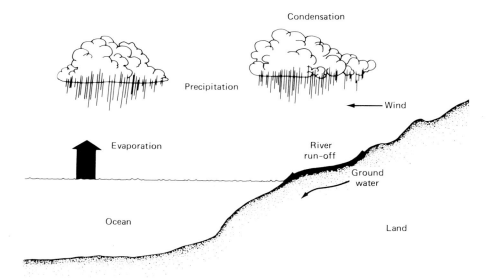

Figure 2-2. The hydrologic cycle. For simplicity, biologic utilization by plants and animals is not shown.

One of the most dramatic series of changes in the ocean has occured over the past million years. I refer to the ice or glacial ages or, as they are known geologically, the Pleistocene Period. Several times during this period large, thick glaciers covered portions of the northern and southern parts of the globe. The most recent of these glacial periods, commonly known as the Wisconsin, started about 30 000 years ago. By 15 000 years ago glaciers had covered all of New England down to Long Island, as well as portions of the interior parts of the United States. An important point is that the water that went into these thick glaciers came from the ocean, with the net result that sea level was 120–130 m lower at the time of maximum glaciation than it presently stands. This lowering of sea level exposed most of the continental shelves of the world, and rivers then deposited their sediments directly onto the continental slope rather than on the shelf. Since 15 000 years ago the sea level has been rising (see Fig. 2-3) and thus submerging the continental shelves. It is not clear whether we are now completely out of this ice age, or whether another advance of the glaciers is imminent.

One important aspect of the recent glaciation is that many portions of the present continental shelf and coastal zone have not yet reached their equilibrium condition. Some environments that we prize highly, such as estuaries and beaches, are still in the natural process of changing and evolving, whereas we, in many instances, are trying to prevent such evolution by dredging or building jetties, etc. This point is discussed further in Chapter 8.

The removal of water from the ocean by glaciation had other effects: for example, it exposed a land bridge between Asia and North America and per-

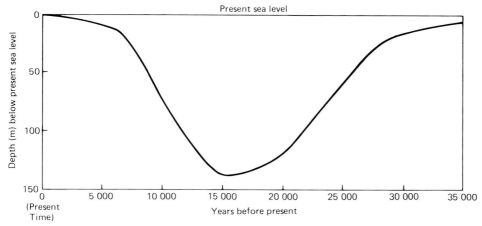

Figure 2-3.   Position of sea level over the past 35 000 years. The shape of this curve has been determined by the collection and dating of items known to form at or near sea level (peat, shells of certain clams, etc.) but now found situated above or below sea level.   In other words, they have been formed in place, but sea level has subsequently risen or fallen, thus "burying" them or exposing them. (Adapted from many sources, including Milliman and Emery 1968)

mitted the migration of animals and perhaps humans (from Asia to North America). Glaciation also must have had some environmental effect in the ocean, because sea water, when it freezes, does not incorporate the salt into its ice structure; the salinity of the ocean therefore must have increased at that time. Likewise, the ocean water temperature must have changed. The effects of these changes are not well known.

General Topography of the Ocean

The ocean has two major topographic features: the continental margin and the deep sea (Fig. 2-4). The continental margin is that portion of the ocean immediately adjacent to land. It in turn can be divided into four main parts: coastal region, continental shelf, continental slope, and continental rise. Probably the two most important parts with respect to uses of the sea are the coastal region and the continental shelf. The coastal region, which is one of the smallest portions of the ocean, is also one of its most transient. The coastal region is that part of the continent adjacent to and directly influenced by the ocean; it includes beaches, marshes, coastal areas, the shoreline, estuaries, and lagoons. Depending on the relief of the coast and the tidal range, the coastal region can range from several hundred feet to several miles in width. Chapter 8 is devoted to the coastal zone.

Continental shelves are that portion of the sea floor adjacent to and sur-

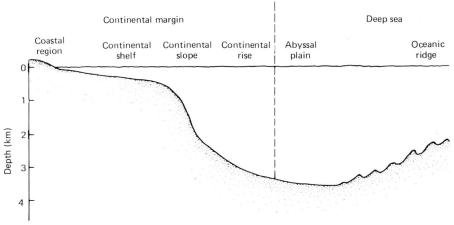

Figure 2-4.   Principal division of the ocean floor: continental margin and deep sea.

rounding the continents or islands. Shelves generally are relatively smooth platforms that end seaward by a fairly abrupt change in slope, called the shelf break, which marks the beginning of the continental slope. The average width of the continental shelf (see Fig. 2-5) is about 40 miles, or about 65 km; however, shelf widths can be over 1000 km. Shelf depths can range from 20 to over 500 m, but the average depth is about 130 m (about 425 ft). Continental shelves cover an extremely large portion of the ocean (Table 2-4) and are equivalent in size to some continents. The point where

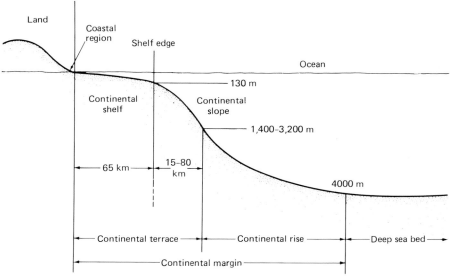

Figure 2-5.   General characteristics of the continental margin.   (Adapted from McKelvey and others 1969)

Table 2-4. Percentage of Major Topographic Features in the Main Oceans and Adjacent Seas[a]

| Oceans and Adjacent Seas | Continental Margin[b] | | Deep Sea | | | |
	Continental Shelves and Slopes	Continental Rise and Partially Filled Sedimentary Basins	Abyssal Plains	Oceanic Ridges	Other Areas	Percent of Total Ocean
Pacific and adjacent seas	13.1	2.7	43.0	35.9	6.3	50.1
Atlantic and adjacent seas	17.7	8.0	39.3	32.3	2.7	26.0
Indian and adjacent seas	9.1	5.7	49.2	30.2	5.8	20.5
Arctic and adjacent seas	68.2	20.8	0	4.2	6.8	3.4
Percent of total ocean in each group	15.3	5.3	41.8	32.7	4.9	100.0

[a]Data from Menard and Smith (1966).
[b]The continental margin has an area of about 74.5 million square kilometers (28.8 million square miles).

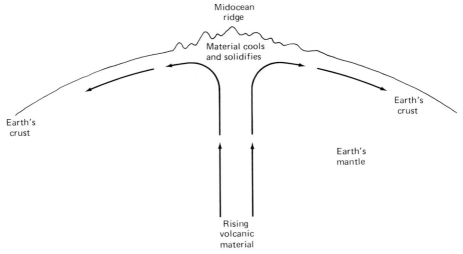

Figure 2-6. The principal aspects of the sea-floor spreading hypothesis. Once the material is solidified, it slowly moves to the right and left, pushed aside and replaced by more rising volcanic material.

the continental shelf ends and the continental slope begins is important for many legal questions and so-called rights of the sea (see Chapt. 3).

Origin of the Ocean Basin

One of the most interesting recent scientfic concepts is that of sea-floor spreading or plate tectonics. This concept (Fig. 2-6) can be used to show how the ocean basins have evolved and changed over geologic time. The process, although not completely understood, is really very simple. New volcanic material rises (by convection from the deep portion of the earth) to the surface, mainly along the midocean ridges that extend across the Atlantic, Pacific, and Indian Oceans (Fig. 2-7). The volcanic material then cools and solidifies, forming a new portion of the sea floor. The process, however, is continuous and the newly formed sea floor is slowly moved aside in a conveyor belt fashion, pushed aside and replaced by still younger material (see also Fig. 5-5). The addition of new material along the ocean ridges occurs at a rate of a few centimeters per year, which, although a low rate, is a considerable one when viewed in terms of the hundreds of millions of years of geologic time. By this process, new sea floor is constantly being added along the ocean ridges and, if it were not for another mechanism, the result of this would be a continuously expanding earth. However, along certain zones of weakness, usually trenches, the now old sea floor collides with and is thrust

Figure 2-7. The general topography of the oceans. Note the long oceanic ridge running north–south along the Atlantic Ocean, which also connects with a ridge in the Indian Ocean and subsequently with one in the Pacific Ocean. The figure is based in part on an excellent series of charts prepared by the late Dr. Bruce C. Heezen and Miss Marie Tharp of the Lamont–Doherty Geological Observatory.

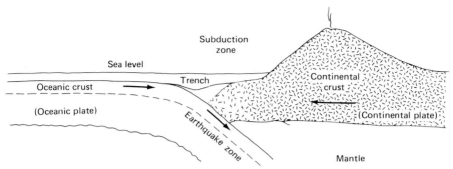

Figure 2-8.   The collision of ocean with continent will generally form a subduction zone, usually with a deep-sea trench. The oceanic plate will be subducted or thrust under the continental plate because of the oceanic plate's heavier density.

under the continent (see Fig. 2-8). These areas are called subduction zones. A more refined aspect of the sea-floor spreading concept is to consider the earth's crust as being composed of a series of plates (see Fig. 2-9). Some parts of plates are in collision with other plates, some plates are sliding along each other, and along some plate edges new crust is being added (along the ridges). Plates containing continental material are generally lighter than oceanic plates, and so oceanic plates generally are thrust under continental plates (Fig. 2-9).

The sea-floor spreading concept has significant implications in the development and discovery of mineral resources (see Fig. 5-5). The concept also allows us to predict both what the positions of the continents and the oceans

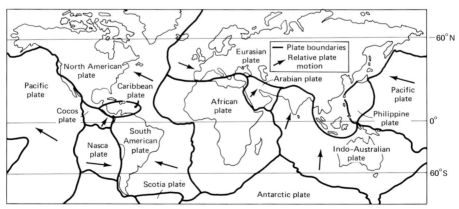

Figure 2-9.   The entire earth's surface can be divided into a series of plates that are moving relative to each other. This figure shows the major plates (some can be further subdivided). The arrows show the movement of the plates relative to the African plate. (Adapted from Ross 1977)

150 to 200 million years ago   80 to 120 million years ago   200 million years in the future

Figure 2-10.   Past and possible future positions of the continents as determined from using the hypothesis of sea-floor spreading.

will be in the future and what they have been in the past (see Fig. 2-10). Probably one of the most appealing aspects of the sea-floor spreading idea is that it can be tested. In other words, there are certain geophysical character-istics of the earth crust that should be evident if the idea is correct—and gen-erally they are. For example, higher than normal heat flow is found along the oceanic ridges, a fact that is predicted by the hypothesis because hot vol-canic material is rising to the surface in this area. Likewise, earthquakes should be, and usually are, common along the ocean ridges and subduction zones because these are areas of crustal weakness. These and other data have given considerable credence to the concept of sea-floor spreading and plate tectonics and help explain many of the features of the sea floor.

Water

Before the different fields of oceanography are discussed, a few words about water are appropriate. There are only three naturally occurring liquids on earth (water, mercury, and petroleum). However, it is only because of a chemical peculiarity that water is a liquid at normal temperatures—an es-pecially important fact, because if water were a gas or solid (ice) at normal temperatures, life as we know it would be impossible. This chemical pecu-liarity is that the water molecule is a "polar" molecule, which means that the electrical charge of the molecule is somewhat unbalanced. The two atoms of hydrogen (which have positive charges) are separated by an angle of 105° when they are attached to an oxygen atom (which has a negative charge). This causes one side of the water molecule to have a negative charge, whereas the other side has a positive charge. (If the two hydrogen atoms were separated by 180° the charges on either side would essentially be equal.) The unequal charge distribution permits water molecules to form clusters with each other instead of existing as individual molecules. More energy (heat for example) has to be applied to break the clustered molecules apart and have them become individual molecules (i.e., to be a gas). If water consisted of single unclustered molecules, it would boil at a temperature of $-80°C$.

The polar property of water is also important in the formation of ice; in this case the water molecules are also bound to each other, but here the relative distance between individual molecules is greater than it is when water is in its liquid form. Because of this, "liquid" water is more dense than ice, and so ice floats in water. This is especially important because if ice were denser than water it would sink, and the deep cold waters of the ocean would be composed of ice. In other words, the ocean would freeze from the bottom up rather than from the top down.

Another important aspect of the polar nature of liquid water is that because of its unequal charge, the water molecule is easily attracted to other elements and can form clusters of water molecules around such elements. This fact explains water's impressive dissolving power. Actually, water is unique in most of its physical properties, and many of these have important implications. One is that it has the highest specific heat of any liquid. This means that water has the ability to store heat energy and that its temperature will rise slower, under a given heat input, than will that of other liquids. One effect of this is that adjacent land and water may often have major differences in their temperature, which in turn can have a considerable influence on weather. The high heat capacity of water also means that it has a large reservoir of heat that could be used as a source of energy (see Chapt. 9).

In the remainder of this chapter, I discuss some aspects of the different disciplines of oceanography. My main goal is to give a brief insight into the principal scientific objectives of physical, chemical, biologic, and geologic oceanography. The reader should remember that oceanography, although commonly divided into individual fields, really involves a broad understanding of several different aspects of science. An example is the use of the coastal zone. It is not satisfactory to know just how a near-shore atomic energy plant may affect fish; it is also important to know how it may affect the currents, water temperature, atmospheric conditions, water chemistry, bottom sediments, plankton, etc., and how these different items, in turn, may interact with each other.

Physical Oceanography

Physical oceanographers are principally interested in the physical properties of the ocean, such as its temperature, salinity, and density, and in determining the reasons for the variations in these properties. One important objective of physical oceanography is to develop a model for predicting ocean water circulation. Studies of large-scale oceanic circulation need numerous measurements, some taken essentially simultaneously over large areas, and some taken over long periods of time. Some physical oceanographers have turned their attention toward studying the distribution of pollutants in the ocean. This can allow time measurements of various ocean phenomena, because it is generally known when (within a few years) the pollutant has been introduced into the ocean. Thus, the presence of a pollutant in the deep water off Antarctica or some other obscure place can indicate the rate of water movement necessary for the pollutant to have gotten there. Another important area of study by physical oceanographers is that of air—sea interactions, because sea water obtains many of its characteristics from the atmosphere. The acoustical properties of sea water are another important area of research, one which also can have military uses (see Chapt. 7).

One way of explaining the physical characteristics of the ocean is to consider what the ocean would look like if it had no motion. If this were the case, temperature would be slightly warmer at the surface than at depth, and salinity would be the same from top to bottom. One of the principal reasons the ocean is not motionless is that whereas all the earth is heated by the sun, this heat is not distributed equally over the earth's surface. More heat is received per unit area at the equator than at the poles because of the curvature of the earth (Fig. 2-11). The result of this is to establish a slow convective movement of the water. Simply said, the water is heated at the equator and moves northward or southward to the poles, where it is cooled, becomes more dense, sinks, and eventually returns to the equatorial region. This type of slow circulation is called thermohaline circulation (*thermo* for heat, *haline* for salt). This circulation is essentially caused by density changes, because water expands and becomes relatively less dense with heating and becomes more dense (and sinks) with cooling. The thermohaline circulation is re-

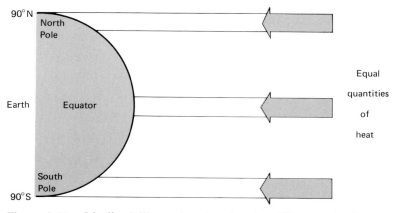

Figure 2-11.   Idealized illustration showing the difference in the amount of heat received per unit area in the polar and equatorial regions.

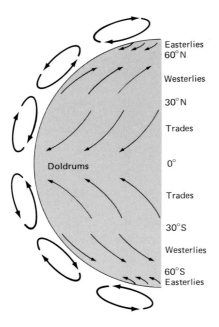

Easterlies
60° N

Westerlies

30° N

Trades

Doldrums

0°

Trades

30° S

Westerlies

60° S
Easterlies

Figure 2-12. Atmospheric circulation and major wind patterns.

sponsible for the deep, slow motion of the ocean. The Pacific ocean is mixed in this manner at a rate of about once every 2000 years or so.

Thermohaline circulation, although very important, is generally obscured in the ocean by the more dynamic wind-driven circulation, which is also a function of the unequal latitudinal distribution of heat received on the earth. The wind-driven circulation is similar to the thermohaline circulation, but here the driving force is the atmosphere. In the wind-driven case, more heat is received by the atmosphere at the equator than at the poles, and the hot air near the equator rises and starts moving north or south toward the pole, where it is cooled (and becomes more dense) and sinks. Because the earth rotates, however, this poleward motion is broken into three gyres, or circulation patterns. These gyres are the major zonal winds that blow on the earth's surface (Fig. 2-12). The atmospheric winds also blow on the ocean and move the surface of the water, eventually developing the wind-driven water circulation pattern (Fig. 2-13). The resulting currents flow not directly in the direction of the wind but at an angle to it (to the right in the northern hemisphere and to the left in the southern hemisphere). The atmosphere, therefore, gets most of its energy from the sun, whereas the ocean gets it both from the sun (thermohaline circulation) and from wind blowing on it (wind-driven circulation). There is more momentum or energy in the wind-driven circulation than in the thermohaline circulation, but the wind-driven circulation is usually restricted to the near-surface layers of the oceans, about the top 1000 or 2000 m. Wind-driven currents can be immense; in the

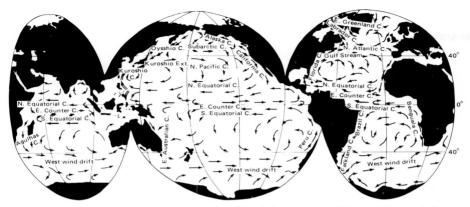

Figure 2-13. The major surface currents (C.) of the ocean, which mainly result from the atmospheric circulation pattern.

Gulf Stream, for example, as much as 100 million cubic meters of water is transported at speeds up to 6 knots along parts of the current.

The blowing of the wind on the ocean surface has several other important effects; two of these are upwelling and waves. Upwelling can result when a wind blows parallel to a coast, which moves surface water away from the coast (Fig. 2-14). This water may be replaced by deeper water that is often enriched in important nutrients. This process leads to a biologic environment very favorable for plant and animal growth. Upwelling is extremely important in the overall biologic economy of the sea, and it is estimated that 50% of the world's fish catch comes from the upwelling zones of the world. It is possible, by using innovative methods, to create artificial upwelling areas (see Chapt. 9, page 287), and in this manner increase biologic productivity.

The other important effect of wind blowing on water is the formation of waves. Waves and their development are an extremely complex subject; however, several aspects are important for the use of the sea, especially in the coastal zone. Further discussion of waves is given in Chapter 8, pages 251–257.

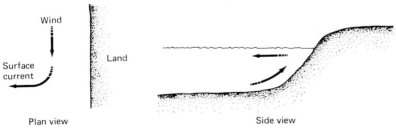

Figure 2-14. Upwelling of near-shore surface waters caused by local or regional winds.

Chemical Oceanography

Simply stated, the chemical oceanographer is principally interested in the distribution of the various elements and compounds in the ocean and the factors that control and influence this distribution. It will be evident from reading the following section on biologic oceanography that many of the factors that influence the distribution of elements are biologic in origin.

A chemical oceanographer needs a water sampling device that obtains enough water for analysis, because many of the components of sea water are present in very small amounts. The sampling device should also be able to be easily and accurately located with respect to depth, temperature, or any other parameter in the ocean and should introduce no contamination (Fig. 2-15). This latter condition is especially important in pollution studies. Present technology has reached the point where there are *in situ* chemical devices that can directly measure certain parameters, such as salinity and oxygen content, and transmit the data back to the ship while the instrument

Figure 2-15. A rosette of water samplers being raised aboard ship. Each of the individual cylinders is a separate water sampler that can be triggered at any depth below the ship. The depth of triggering is determined by electronic devices that monitor and transmit to the ship data on various properties of the water, such as its temperature. At a particularly interesting point a water sample can then be taken. (Photograph courtesy of Woods Hole Oceanographic Institution)

Table 2-5. Major Components of Sea Water[a]

Element	Concentration (ppm)
Chlorine	19 000
Sodium	10 500
Magnesium	1 350
Sulfate	885
Calcium	400
Potassium	380
Bromine	65
Carbon	28
Strontium	8
Boron	4.6

[a]At least 60 other elements are present in quantities of less than 1 ppm.

is still in the water. Methods of analysis are highly sophisticated and are common to those used in many modern labs.

Oceanographers, when they talk about the salt content of sea water, use the term "salinity," which is a measure of the total dissolved inorganic compounds in the water. Generally, sea water has a salinity of 35 parts per thousand (35‰ or 3.5%). The common elements in sea water are listed in Table 2-5.

One of the most important principles in chemical oceanography is that although there can be marked differences in the total salinity between different samples of sea water, the ratio of the major constituents in the samples has been discovered to be invariant or constant. For instance, the ratio of calcium to magnesium, or of magnesium to chlorine, will essentially stay the same, regardless of the total amount of these elements present in the sea water. This principle is a powerful tool, because by measuring only a single component of sea water, one can essentially determine the general composition of the sample. Unfortunately, there are other elements in sea water, particularly such nutrients as nitrogen, phosphorus, and silicon, that will vary considerably in their concentrations because of biologic reactions. These variations occur both on a seasonal basis and with depth in the water column. The elements that usually have a constant ratio in sea water are called *conservative elements*; ones that are involved in biologic reactions are called *nonconservative elements*.

Gas content of sea water can also change because of biologic reactions, especially photosynthesis. Because of this reaction, oxygen content is high in the surface waters at noon, when photosynthesis is at a maximum (because sunlight is at a maximum), and low at night (no sunlight), when the opposite of photosynthesis—respiration—prevails. Photosynthesis (the production of organic matter by plants) is easily one of the most important reactions in the ocean; it is dependent on light and the presence of nutrients and can be expressed as:

$$H_2O + CO_2 \xrightarrow{\text{Light and nutrients}} CH_2O + O_2 \quad \text{(Photosynthesis)}$$

where CH_2O is a simplified description of organic matter. In this reaction, oxygen and organic matter are produced and water and carbon dioxide are consumed. When the reaction goes in the other direction:

$$H_2O + CO_2 \longleftarrow CH_2O + O_2 \qquad \text{(Respiration)}$$

it is called respiration, or oxidation, and oxygen and organic matter are consumed and water and carbon dioxide are produced. Photosynthesis, because it is dependent on sunlight, can only occur in the uppermost layers of the ocean, usually the upper 100 or 200 m. Respiration can occur anywhere in the ocean. Photosynthesis in the ocean is mainly done by floating plants, called plankton. By the process of photosynthesis, many elements (nutrients) are removed from the surface waters and incorporated into organic matter (plants). If this organic matter is consumed by animals and is eventually excreted, or if the original plant or consuming animal dies and settles down through the water and is oxidized, the nutrients are returned to the water, although usually at a depth greater than they were consumed. In time these nutrients may be returned to the surface waters, usually by upwelling (see Fig. 2-14), and be recycled again by photosynthesis. The process whereby nutrients are removed from the surface waters and transported to depths and then returned to the surface waters is an extremely critical one for the biologic and chemical aspects of the ocean. Without this mechanism of recycling, life in the oceans would be reduced considerably. Production of organic matter by plants, the basic source of food in the ocean, can only take place if the supply of nutrients is adequate. If the water is not mixed, the nutrients that have settled in the deeper layers remain there, not to be available for photosynthesis in the surface waters (where the light and therefore the plants are).

A typical distribution of some parameters of sea water with depth is shown in Fig. 2-16. Temperature generally is at a maximum in the surface layers and decreases fairly rapidly until a depth of about a kilometer or so, where there is then a slow decline toward the bottom. The zone of most rapid temperature decrease is called the *thermocline* and separates the upper and lower parts of the ocean by density differences. In other words, the surface waters are relatively light because they are warm, and the deep waters are relatively heavy because they are cold. The thermocline is the zone of most rapid temperature change and so the area of the largest change in den-

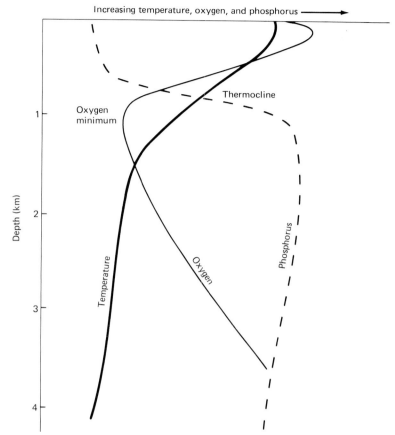

Increasing temperature, oxygen, and phosphorus ⟶

Figure 2-16. Distribution of temperature, oxygen, and phosphorus (a nutrient) with depth.

sity. The higher the density difference, the more stratified the water. Therefore, the thermocline tends to prevent or retard the vertical mixing of water; the thermocline has to be overcome or destroyed before the water can be thoroughly mixed; upwelling is one such mechanism of mixing.

Oxygen in sea water has a distribution somewhat similar to that of temperature, in that it is high in the surface waters. This high concentration results from two key processes: photosynthesis, which occurs in surface waters and produces oxygen, and exchange or mixing with the atmosphere. Oxygen decreases fairly rapidly with depth as photosynthesis decreases (because of diminishing light) and respiration predominates. During respiration, oxygen is consumed. Respiration, of course, also occurs where photosynthesis is high, but here the relative rate of respiration is small and so there is a net production of oxygen. The increase in oxygen with greater depth results from the thermohaline circulation (see pages 19–20) where surface waters, rich in oxygen, have sunk in the polar regions and slowly move equatorward along the bottom.

In some areas the consumption of oxygen at intermediate depths can be so high as to remove essentially all the oxygen from the water, resulting in what is called a reducing or anaerobic condition (this layer is called the oxygen minimum zone). The oxygen minimum zone can be prevalent in closed basins or areas of restricted circulation, such as the Black Sea or the Cariaco Trench. Both of these areas are isolated from a source of deep water that could replenish their oxygen supplies, and the high production of organic matter in the overlying waters, combined with the subsequent decay or respiration of this material, has used up all the oxygen, resulting in an extensive oxygen minimum zone. There may be an economic importance to this process because organic matter settling into an oxygen minimum zone (or falling at a rate faster than it can be decayed) can accumulate on the bottom instead of being decomposed. If this organic matter accumulates, is buried by sediments, and undergoes the appropriate chemical reactions it can, over a few million years, eventually become a petroleum deposit. In fact, it is this very mechanism occurring in the marine environment that is responsible for many of our present oil deposits.

The vertical distribution of a typical nutrient, for example phosphorus (Fig. 2-16), is like that of oxygen—essentially controlled by photosynthesis. Phosphorus is low in the surface waters because it has in most likelihood been involved in photosynthesis and incorporated into organic matter and so has been effectively removed from the water. Its content, however, increases in the deeper layers because it is removed by respiration from the decaying organic matter and returned to the water. If the phosphorus is moved back to the surface waters (where there is adequate sunlight) it can again get involved in the photosynthetic process.

The above is a simplified picture of a very complex process than can vary with the season, among other things. For example, in early spring and fall there may be periods of high biologic productivity, commonly called blooms. At these times the nutrient content will almost be depleted from the surface waters because of the rapid production of organic matter. In the winter time, however, upwelling and mixing of the waters by storms can replenish the nutrient supply in the surface water; usually, however, there is not enough sunlight for photosynthesis to proceed at a rapid rate. A simplified picture of the distribution and movements of the nutrient phosphate is shown in Fig. 2-17. As previously stated there are certain times of the year when a larger percentage of the nutrient is in the plants, whereas at other times it may be in solution or elsewhere.

One of the more interesting aspects of the different elements in sea water is that those that are very common on land are not particularly the ones that are most abundant or even common in the ocean. For example, iron, aluminum, and silicon are very abundant on land but are not common in the

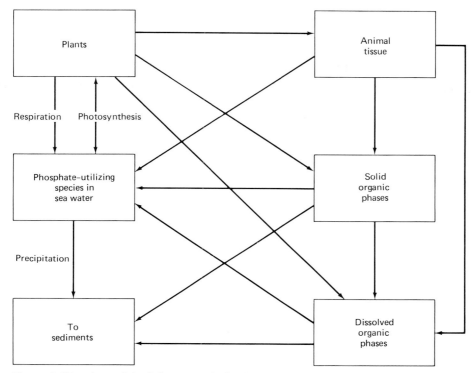

Figure 2-17.   A model of the general distribution of the nutrient phosphate in sea water and in the biologic cycle.

ocean; whereas chlorine and sodium are fairly rare on land but very abundant in the ocean. This apparent contradiction (because most of the elements present in the ocean have been derived from land) results from differences in the reactive abilities of the elements in sea water. The very reactive elements are removed (either biologically or chemically to the sediments), whereas the unreactive elements remain in the water and become relatively enriched.

Biological Oceanography

Biologic oceanographers are principally concerned with determining the distribution of the animals and plants in the ocean and with understanding the interrelationships of these organisms with the marine environment and with each other. To accomplish this, the biologic oceanographer needs a method of sampling the organisms in the ocean. This is not as easy as it sounds, because many creatures, such as fish, can swim and easily avoid most sampling devices. Even more important is the fact that most organisms in the sea tend to be distributed in clumps or groups. This means that individuals are not usually found isolated—one at a time—but tend to occur together in large numbers. It is possible to tow two sampling devices simultaneously through the water, separated by only a few meters distance, and have one device obtain literally thousands of one kind of microscopic creature, while the other will only obtain a few. This clumped distribution complicates most methods of statistical analysis and is especially bothersome to fishery biologists, who are marine biologists specializing in the study of fish and their distribution.

Most sampling techniques involve either pulling something through the water—usually a net—or dragging something along the bottom. A common instrument is the plankton net, a device that can be towed either horizontally through the water while the ship is moving or in a vertical direction when the ship is stopped (Fig. 2-18). It can be made to close or open at selected depths, but in any case it generally can catch only slow-moving creatures. Some organisms tend to live in the surface waters during the day and in

Figure 2-18. Two large plankton nets being prepared for towing. (Photograph courtesy of Woods Hole Oceanographic Institution)

deeper waters at night, or vice versa; the time of sampling therefore can determine what is caught. Other types of sampling devices and methods include dredges, scuba diving, cameras, and electronic devices, such as echo sounding equipment that can detect organisms by the reflection of sound off them (see Fig. 5-26). Fishermen and fish-tagging operations can also provide considerable information about distribution and movement of organisms. Many marine fishery programs involve collecting data routinely from distinct areas over long periods of time. By analysis of these types of data, assessment of the fishery stock is often possible.

A key objective of sampling and other biologic measurements is to determine the rate of the production of organic matter, which is a reflection of biologic activity. Biologic relationships in the ocean are somewhat different than those on land. For example, the animals in the ocean are in much greater intimacy with their environment than are those on land. Therefore, any major change in the marine environment, such as variations in temperature and salinity, will have a very quick and dramatic effect on the organism. In the open ocean many marine animals and plants live at near lethal conditions and are especially vulnerable to change. El Niño, one of these changes, is discussed on page 293. Organisms that live in the near-shore or coastal environment, however, can tolerate a more variable environment with greater changes.

Probably the most important biologic difference between the ocean and land is that only a very small portion of the sea floor is available for plant growth, because plants have to live in shallow depths so that they can receive sufficient sunlight to grow. Probably less than 1% of the sea floor is shallow enough and also has the right type of bottom for plant growth. Because of this, the plants in the ocean are mainly phytoplankton, a term for microscopic organisms that float in the ocean (Fig. 2-19). It is these creatures that produce most of the primary organic matter in the ocean (by photosynthesis).

In general, there are very large ranges in many of the major characteristics of the ocean. Salinity can vary from 0 to 40 parts per thousand (‰), the high values being common in such places as the Red Sea, whereas low values can occur in estuaries. Temperatures can be less than 0°C (below 32°F) in the deep ocean and as high as 35°C (almost 100°F) in tropical areas and such places as the Persian Gulf. Water depth ranges from the surface (0) to over 11 000 m, and this creates pressure ranges of from 1 atm to over 1000 atm. Essentially, 10 m of depth equal 1 atm of pressure. Light varies from bright sunlight on the surface waters to complete darkness at depth. In spite of these tremendous variations, there are large areas of the oceans that have essentially uniform environmental conditions (see Fig. 2-20), and no animal or plant has to bridge such a wide range of environments.

The organisms that live in the ocean can be classified into three main groups. *Benthos* are bottom-living animals, such as clams and worms, and

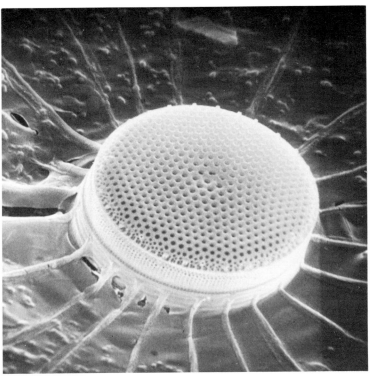

a

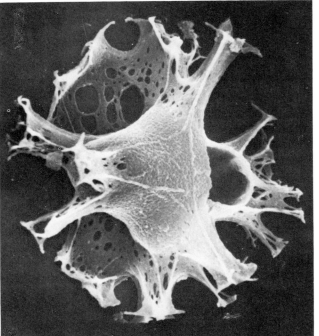

Figure 2-19, a and b. Electron microscope photograph of two common plankton: **a** a diatom and **b** a dinoflagellate. Both creatures are about $\frac{1}{500}$ cm in diameter. (Photographs courtesy of Dr. Susumu Honjo)

b

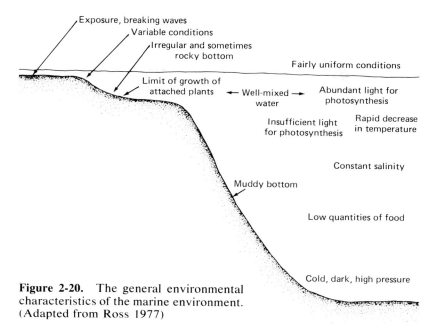

Exposure, breaking waves

Variable conditions

Irregular and sometimes
rocky bottom

Fairly uniform conditions

Limit of growth of
attached plants

← Well-mixed →
water

Abundant light for
photosynthesis

Insufficient light
for photosynthesis

Rapid decrease
in temperature

Constant salinity

Muddy bottom

Low quantities of food

Cold, dark, high pressure

Figure 2-20. The general environmental characteristics of the marine environment. (Adapted from Ross 1977)

bottom plants (in shallow water only); *Nekton* are swimming organisms, which may live anywhere but usually stay near the surface or near the bottom; *Plankton* are floating organisms that can be either phytoplankton (which are plants) or zooplankton (which are animals). It should be emphasized that animals can go from one of these main groups to another during their life histories; in fact, many marine animals spend the early portion of their lives as plankton and then become nekton or benthos. Zooplankton generally eat microscopic phytoplankton (diatoms are a common variety) and, in turn, zooplankton are eaten by animals higher up in the food chain.

The key ingredient in the biologic economy of the ocean is the production of organic matter (Fig. 2-21) by plants (commonly called primary production). It is usually described as the amount of organic matter produced per unit area or per unit volume for a given period of time. The plants (i.e., the organic matter) eventually die or are consumed by animals higher in the food chain—usually zooplankton. In the latter case, the plant material becomes part of the animal's organic matter (commonly called secondary production). Eventually this animal either dies or is consumed and the process continues. When the animals or plants die and decay, their organic matter and associated nutrients, etc., return to the water, perhaps to be recycled by upwelling or other processes or perhaps to become incorporated into the bottom sediments.

Most of the biologic reactions in the ocean can be considered as part of a food chain process. A simple model is shown in Fig. 2-22, where phy-

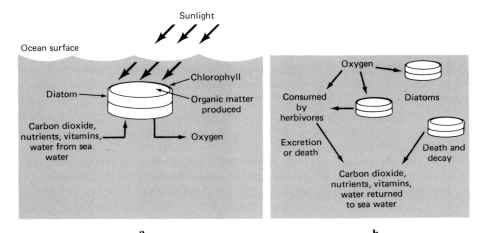

Figure 2-21, a and b.   The basic productivity cycle in the ocean. **a** Photosynthesis and **b** respiration. (Adapted from Ross 1977)

toplankton are at the bottom and large predators at the top. The conversion ratio of food from one level to another is small, perhaps in the range of 10%; i.e., 1000 lb of phytoplankton will yield 100 lb of zooplankton and, eventually, 1 lb of tuna. The question of efficiency of conversion of organic matter is discussed further in Chapter 5. Perhaps a more realistic view of the biologic process is illustrated in Fig. 2-23, which shows the idealized pathways of nutrients within the biologic environment.

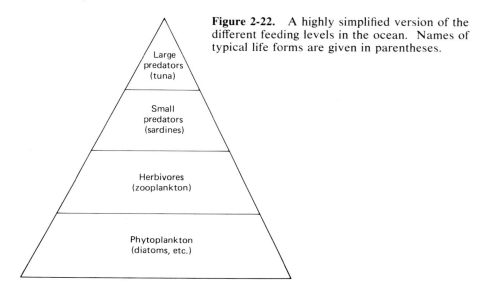

Figure 2-22.   A highly simplified version of the different feeding levels in the ocean. Names of typical life forms are given in parentheses.

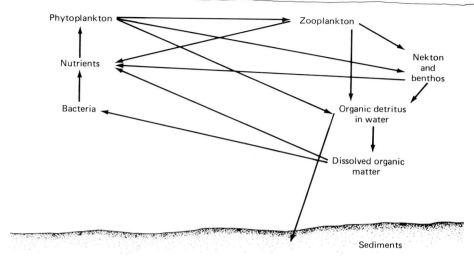

Figure 2-23.   The life cycle in the ocean. (Adapted from Raymont 1963)

Marine Geology–Geological Oceanography

The marine geologist is principally concerned with the ocean bottom and its sediments, as well as with the history of these features and the processes that have caused them to have their particular characteristics.   A marine geophysicist is more interested in the deep structure of the ocean basin and its physical characteristics. Basic equipment to both of these fields includes echo sounders for determining the depth (bathymetry) of the ocean, cameras to photograph the sea floor (Fig. 2-24), and sampling devices, such as a corer, grab, or dredge. Geophysical equipment includes magnetometers (for measuring the earth's magnetic field), gravimeters (for measuring the earth's gravitational field), and seismic profilers. This latter piece of equipment deserves a further description because of its importance in locating areas having oil potential. A seismic profiling device (the technique is commonly called continuous seismic profiling, or CSP) emits large quantities of acoustical energy into the water (Fig. 2-25). The energy travels in all directions, and some reaches the sea floor and is reflected back to the ship, similar to the process of echosounding.   Some energy, however, penetrates to layers below the sea floor and is also reflected back, generally from layers that have distinct changes in their acoustical characteristics.   These acoustical changes usually indicate different sediment or rock types. By receiving the returning signals and using computer techniques, it is possible to determine the structure and seismic velocities of the upper layers of the earth's crust.

a

b

Figure 2-24, a and b.   **a** Lowering of a deep-sea camera system.   **b** A somewhat untypical bottom photograph showing a large group of holothurians (sea cucumbers) feeding on the continental slope. (Photograph courtesy of Woods Hole Oceanographic Institution).

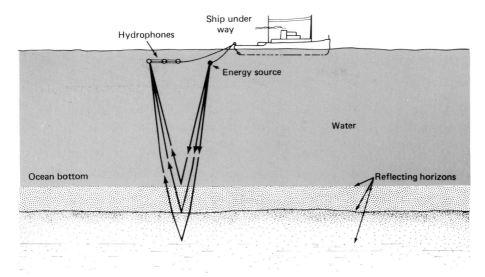

Figure 2-25. The geophysical technique of continuous seismic profiling. Sound energy is reflected back and received from layers having different acoustic properties. See Figs. 5-12 and 5-14 for actual records.

These techniques are commonly used in surveys to determine the potential for oil and gas (although drilling is always necessary to verify the deposit).

The crustal structure beneath the deep sea and beneath the continent is different, and the continental margin is the transition area between the two (see Fig. 2-4). The crust is considerably thicker under the continents and is composed mainly of granitic rock, whereas basaltic rocks are typical under the ocean basin. The origin of the ocean has been discussed previously (see pages 14–17), and the concept of sea-floor spreading and plate tectonics appears to answer most of our questions concerning the major features of the sea floor.

The continental margin generally can be divided into a continental rise, a continental slope, a continental shelf, and a coastal region (see Fig. 2-5); the total area of the continental margin is over 20% of the total ocean (over 72 million square kilometers). Continental shelves have been fairly well studied only in a few areas of the world and are the site of most of the biologic and mineral resources of the ocean (see Chapt. 5 for details). The continental margin is an especially important area in the ongoing discussions concerning the Law of the Sea (see Chapt. 3), because in many instances much of the margin (especially the continental shelf and slope) falls within the proposed 200-mile economic resource zone.

Continental rises, which cover a large portion of the ocean (see Fig. 5-12), generally have thick sequences of sediments that may have oil and gas poten-

tial. Technology is not sufficient to exploit this area at the present time;
however, it may develop within a few years when such resources are found.

The near-shore parts of the continental margin, commonly called the
coastal zone, are an especially important area of the sea. It is in this region
that competition for use is most intense and complex. For example, should
coastal areas be developed for commercial or for private use or be set aside
for conservation, etc.? The subject of the coastal zone is discussed further in
Chapter 8.

The deep sea has varied importance, seen from the viewpoint of the uses
of the sea. It is the arena for much military (Chapt. 7) and shipping operations (Chapt. 4), but aside from manganese nodules (Chapt. 5), it probably is
not too important for mineral or biologic resources. The following chapters
of this book will further expand on the various aspects of the ocean mentioned in this chapter.

References

Broecker, W.A., 1974, *Chemical Oceanography*. New York: Harcourt, Brace,
    Janovich. 214p

Bullard, E., 1969, The origins of the oceans: *Sci. Amer.*, v. 221, n. 3, p. 66–75.

Deacon, G.E.R., 1962, *Seas, Maps, and Men: An Atlas-History of Man's Explora-
    tion of the Oceans*. Garden City, N.Y.: Doubleday.

Friedrich, H., 1970, *Marine Biology: An Introduction to Its Problems and Results*.
    Seattle: Univ. Washington Press.

Gross, M.G., 1972, *Oceanography: A View of the Earth*. Englewood Cliffs, N.J.:
    Prentice-Hall, 581pp.

Harvey, H.W., 1960, *The Chemistry and Fertility of Sea Waters*. Cambridge,
    England: Cambridge Univ. Press.

Heezen, B.C., and C.D. Hollister, 1971, *The Face of the Deep*. New York: Oxford
    Univ. Press.

Holt, S.J., 1969, The food resources of the ocean: *Sci. Amer.* v. 221, n. 3, p.
    178–194.

*Leopold, L.P., and K.S. Davis, 1966, *Water*. New York: Time Incorporated.

MacIntyre, F., 1970, Why the sea is salt: *Sci. Amer.* v. 223, n. 5, p. 104–115.

McConnaughey, B.N., 1970, *Introduction to Marine Biology*. St. Louis: C. V.
    Mosby.

*Menard, H.W., and S.M. Smith, 1966, Hypsometry of ocean basin provinces:
    Jour. Geophys. Res., v. 71, p. 4305–4325.

*Milliman, J.D., and K.O. Emery, 1968, Sea levels during the past 35,000 years:
    Science, v. 162, p. 1121–1123.

Processes. Oxford, England: Pergamon Press.

Penman, H.L., 1970, The water cycle: *Sci. Amer.*, vol. 223, n. 3, p. 99–108.

Pirie, R.G., 1973, *Oceanography: Contemporary Readings in Ocean Sciences*.
    New York: Oxford Univ. Press,

*References marked with an astrisk are cited in the text; others are of a general nature.

*Raymont, J.E.G., 1963, *Plankton and Productivity in the Oceans.* New York: Pergamon Press.

Riley, J.P., and R. Chester, 1971, *Introduction to Marine Chemistry.* London: Academic Press.

Riley, J.P., and G. Skirrow (Eds.), *Chemical Oceanography* (a continuing series of volumes). New York: Academic Press.

*Ross, D.A., 1977, *Introduction to Oceanography,* 2nd ed. Englewood Cliffs, N.J.: Prentice-Hall, 438 pp.

Schlee, S., 1973, *The Edge of an Unfamiliar World: A History of Oceanography.* New York: E.P. Dutton and Company.

Shepard, F.P., 1959, *The Earth Beneath the Sea.* Baltimore: Johns Hopkins Univ. Press.

Shepard, F.P., 1973, *Submarine Geology.* New York: Harper & Row.

Skinner, B.J., and K.K. Turekian, 1973, *Man and the Ocean.* Englewood Cliffs, N.J.: Prentice-Hall.

Spar, J., 1973, *Earth, Sea, and Air: A Survey of the Geophysical Sciences.* Reading, Mass.: Addison-Wesley.

Sverdrup, H.U., M.W. Johnson, and R.H. Fleming, 1942, *The Oceans: Their Physics, Chemistry and General Biology.* Englewood Cliffs, N.J.: Prentice-Hall.

Tait, R.V., and R.S. DeSanto, 1972, *Elements of Marine Biology.* New York: Springer-Verlag.

Teal, J., and M. Teal, 1969, *Life and Death of the Salt Marsh.* Boston: Little, Brown.

The Economic Value of Ocean Resources to the United States. Report prepared for the use of the Committee on Commerce, National Ocean Policy Study, 1974, 109 pp.

von Arx, W.S., 1975, *An Introduction to Physical Oceanography.* Reading, Mass.: Addison-Wesley.

Weyl, P.K., 1970, *Oceanography: An Introduction to the Marine Environment.* New York: J. Wiley & Sons.

Wilson, J.T., 1972, Introduction: *In: Continents Adrift: Readings from Scientific American.* San Francisco: W.H. Freeman.

General Oceanography Texts

Anikouchine, W.A., and R.W. Sternberg, 1973, *The World Ocean: An Introduction to Oceanography.* Englewood Cliffs, N.J.: Prentice-Hall, 338 pp.

Davis, R.A. Jr., 1972, *Principles of Oceanography.* Reading, Mass.: Addison-Wesley, 434 pp.

Duxbury, A.C., 1971, *The Earth and Its Oceans.* Reading, Mass.: Addison-Wesley, 381 pp.

Gordon, B.L. (Ed.), 1970, *Man and the Sea.* Garden City, N.Y.: Natural History Press, 498 pp.

Gross, M.G., 1972, *Oceanography: A View of the Earth,* Englewood Cliffs, N.J.: Prentice-Hall, 581 pp.

Ross, D.A., 1977, *Introduction to Oceanography,* 2nd ed. Englewood Cliffs, N.J.: Prentice-Hall, 438 pp.

Chapter 3

The Legal Aspects of the Ocean

For centuries it was thought that the oceans were so immense that their resources and capacity to receive waste products from land were inexhaustible. As the use of the sea has increased, this assumption has been found to be incorrect. This realization, combined with the finding of new and important marine resources and the development of the technology to get these resources, has started a new kind of "land rush," this time in the ocean. Before this decade is over it is probable that the countries of the world will have obtained some form of national jurisdiction over at least some 37 745 000 square nautical miles (over 100 million square kilometers) of the ocean floor. This is an area equivalent to 87% of the land area of the earth. These new territorial and economic claims have been discussed and debated in a series of Law of the Sea Conferences which started in late 1973. This chapter describes the events leading up to these conferences and some of the results.

Events Leading to the Truman Proclamation

Regulations and certain freedoms of the sea have been an established marine tradition almost since the time that we first started to sail the ocean. Certainly, early sailors had an unwritten code of conduct concerning sea-going operations, accidents at sea, and navigation that had developed either out of necessity or from custom. Likewise, the strongest country or navy generally imposed their will on less well-endowed nations. Then, perhaps as now, the customary or traditional law of the sea favored those developed countries.

The question of who owns the oceans had been asked and answered several times in the past. The Pope, in 1493, issued a Papal Bull that divided the oceans between the two major ocean powers of that time—Spain and Portugal. This decision gave these countries exclusive trading privileges with the East and West Indies, but the privilege lasted only as long as Spain and Portugal had the power to enforce their "rights." In 1588, England and the Netherlands defeated the Spanish Armada and opened a new era of trade for most countries.

Credit for the first significant legal statement concerning the law of the sea is usually given to a Dutch lawyer, Hugo Grotius. In 1609, he published his *Mare Liberum*, or the concept of the freedom of the seas. Grotius' declaration stated that nations are essentially free to use the sea for whatever purpose they wish as long as this use does not interfere with another nation's use of the sea. It was realized by this time that a coastal state should have some control over the ocean immediately adjacent to its coast. Therefore a zone, called the territorial sea, was established, over which the coastal nation had complete sovereignty. This zone came from a ruling by the Dutch jurist Bijnkershoek that a state could claim territory in a seaward direction for as far as a cannon could fire. The territorial sea was generally accepted as being 3 miles in width, based on the effective distance that a cannon of that time could reach. Thus, in a way, the width of the territorial sea was determined by the technology of the time. Not all nations accepted this 3-mile width and it was never officially ratified by an international agreement. It is interesting to note that as technology improved, i.e., as cannons were built that could shoot farther, there was no strong attempt, until very recently, to widen the territorial sea.

The area outside of the territorial sea, the high sea, was considered as *res nullius*, or belonging to no one. Among the freedoms of the high sea are those of navigation, fishing, overflight, the laying of pipelines and cables (formalized in the 1958 Geneva Convention on the High Seas), mining of natural resources, and military operations. Scientific research without regulation also was considered a basic freedom of the high seas. In general, international law protects the rights of countries to use all parts of the high seas in essentially whatever manner or purpose they choose. This concept was not significantly challenged prior to World War II because the interests of most of the countries that were using the sea (during peacetime) tended to coincide. However, after World War II, the development of many new nations, the decolonization of others, the general awareness that certain countries were more developed than others, and some actions by the United States led to convergent interests in the sea and the beginning of challenges to the previously accepted freedoms and laws of the sea. Another contributing aspect was the growing awareness that some of the resources of the sea, such as fish, were not inexhaustible, whereas others, such as minerals, might be extremely valuable and could be of considerable importance to a developing country.

The Truman Proclamation

One of the first major important challenges to the 3-mile territorial sea concept was made by United States President Truman in 1945, in what is now called the Truman Proclamations on the Continental Shelf. Actually, two distinct policy statements were made. The first established a national policy with respect to the natural resources of the subsoil and sea bed. It said that the United States felt:

> that the exercise of jurisdiction over the natural resources of the subsoil and sea bed of the continental shelf by the contiguous nation is reasonable and just, since the effectiveness of measures to utilize or conserve these resources would be contingent upon cooperation and protection from the shore, since the continental shelf may be regarded as an extension of the land-mass of the coastal nation and thus naturally appurtenant to it, since these resources frequently form a seaward extension of a pool or deposit lying within the territory, and since self-protection compels the coastal nation to keep close watch over activities off its shores which are of the nature necessary for utilization of these resources.[1]

Thus "the Government of the United States regards the natural resources of the subsoil and sea bed of the continental shelf beneath the high seas but contiguous to the coasts of the United States as appertaining to the United States, subject to its jurisdiction and control."[1] One of the motivations for such an action was the "long-range world-wide need for new sources of petroleum and other minerals."[1]

A second proclamation was issued to distinguish between the resources of the sea bed and subsoil (legal jargon for the ocean floor and underlying sediments and rock) and the resources of the overlying water. This stated that:

> in view of the pressing need for conservation and protection of fishery resources, the Government of the United States regards it as proper to establish conservation zones in those areas of the high seas contiguous to the coasts of the United States wherein fishing activities have been or in the future may be developed and maintained on a substantial scale. Where such activities have been or shall hereafter be developed and maintained by its nationals alone, the United States regards it as proper to establish explicitly bounded conservation zones in which fishing activities shall be subject to the regulation and control of the United States. Where such activities have been or shall hereafter be legitimately developed and maintained jointly by nationals of the United States and nationals of other States, explicitly bounded conservation zones may be established under agreements between the United States and such other States; and all fishing activities in such zones shall be subject to regulation and control as provided in such agreements. The right of any State to establish conservation zones off its shores in accordance with the above principles is conceded, provided that the corresponding recognition is given to any fishing interests of nationals of the United States

[1]*Policy of the United States with Respect to the Natural Resources of the Subsoil and Seabed of the Continental Shelf,* Proclamation No. 2667 (Washington, D.C.: U.S. Government Printing Office, 1945).

which may exist in such areas. The character as high seas of the areas in which such conservation zones are established and the right to their free and unimpeded navigation are in no way thus affected.[2]

Essentially, Truman was claiming ownership for the United States of all sea-bed resources on or under the continental shelf and suggesting that some sort of international agreement be reached concerning regulation of fisheries and conservation. The proclamation led to considerable controversy and had numerous unanticipated results. Lost in much of the discussions was that the United States was not really changing its 3-mile territorial sea. The proclamation, in addition, was not implemented in any way to exclude, for example, foreign fishing outside of 3 miles off the coast. However, in claiming ownership of parts of the sea floor, but not of the overlying waters, it had to influence some of the basic freedoms of the overlying sea.

The Truman Proclamation also caused conflicts between the coastal states of the United States and the Federal Government concerning offshore ownership and mineral claims. Prior to 1945 it was generally thought that each of the individual states of the United States had title to the lands immediately off their coasts. Oil exploration began off California in the 1890s without any Federal control and initially without even any state control. The Truman Proclamation, however, claimed Federal ownership of sea-bed mineral resources and thus started a series of Supreme Court cases between various states and the Federal Government. In 1947, the Supreme Court (*United States* vs. *California*, 332, U.S. 19) said that the Federal Government had full dominion and power over the lands, minerals, and other resources underlying the Pacific Ocean lying seaward of the ordinary low-water mark on the coast of California and outside of the inland waters. Subsequent cases between the United States vs. Texas and the United States vs. Louisiana also gave the Federal Government ownership of the submerged land off these states.

The Court's actions were changed by two Congressional acts passed in 1953. The first, the Submerged Lands Act, gave the states title to the lands and resources out to a distance of 3 geographic miles in the Atlantic and Pacific Oceans and out to 3 leagues or 10.5 statute miles in the Gulf of Mexico. The Federal Government retained within the zone its rights concerning navigation, commerce, national defense, and international affairs. This act thus defined the *inner continental shelf* as extending from the low-water line to the end of the 3-mile territorial sea. The *outer continental shelf* is the rest of the continental shelf seaward of the territorial waters. The second act, the Outer Continental Shelf Lands Act, gave control of the sea bed and subsoil of the outer continental shelf to the Federal Government.

After these acts were passed and declared constitutional, several gulf states challenged the width of their title, based in part on historical claims

---

[2]*Policy of the United States with Respect to Coastal Fisheries in Certain Areas of the High Seas*, Proclamation No. 2668 (U.S. Government Printing Office, 1945).

when they joined the Union or on whether their claims should start from their outer islands or from the coast. The result was that western Florida and Texas were given boundaries that extended 9 miles from the coast, whereas Louisiana, Mississippi, and Alabama got only 3 miles. Several later court cases ruled against states' starting their claims other than from mean lower low water and not allowing them to extend their borders seaward by artificial changes in their coast.

More recently, several east coast states began litigation to obtain further property rights into the Atlantic Ocean. The states based their claims on colonial charters and early history, in some instances claiming rights to as far as 100 miles off their shore. Eventually, 13 states became part of the legal action. In 1969, Maine tried to authorize private exploitation of the sea floor in the Gulf of Maine, parts of which were about 88 miles off the coast. In August 1974, a Special Master on behalf of the United States Supreme Court rejected the states' claims on the basis of previous court decisions giving ownership to the states out to 3 miles. The decision also noted that any rights that the early colonies had, would have been passed onto the Federal Government when they became members of the United States. The Supreme Court can still consider the question further. The case is of considerable financial importance to these states, and to others, because of the oil and gas potential on the outer parts of the continental shelf. For the northeastern coastal states there is little probability of oil and gas within the 3 miles immediately off the coasts that they now own.

The 1958 and 1960 Geneva Conventions on the Law of the Sea

The Truman Proclamation actually did not define the continental shelf, although a press statement released by the White House on the same day as the Proclamation said "generally, submerged land which is contiguous to the continent and which is covered by no more than 100 fathoms (600 feet) of water is considered as the continental shelf." The same press release said that the Truman Proclamation does not "abridge the right of free and unimpeded navigation of waters of the character of high seas above the shelf, nor does it extend the present limits of the territorial waters of the United States." However, the challenge to the existing regime had been made and the disclaimers in a press release were not very effective.

These claims by the United States were treated with suspicion by some countries, with indifference by others, and as an opportunity to adust supposed injustices by still other countries. Among the latter were several South American countries that had few offshore resources and were anxious to protect what they had. In 1952, Chile, Ecuador, and Peru signed a

Declaration on the Maritime Zone (better known as the Declaration of San-tiago),[3] which extended their jurisdiction and sovereignty over the sea, sea bed, and subsoil out to a distance of 200 miles from their coastlines. This was essentially declaring a 200-mile territorial sea. It differed considerably from the Truman Proclamation in that it covered the sea surface and water, as well as the ocean bottom, and extended far beyond any reasonable es-timate of their countries' continental shelves. Peru argued that it had an im-portant industry—guano, used for fertilizer—on islands off its coast which was extremely important for their economy and should be under their con-trol. In later years, with help from the United States and the United Na-tions, they developed a major anchovy fishery which, for most of the 1960s, made Peru one of the world leaders in fish catch. American boats caught fishing within the 200-mile limits of their coasts have been seized and even-tually released when a fine (usually in the tens of thousands of dollars) is paid. The United States Government, which does not recognize these claims, will reimburse the owner of the fishing boat. The situation has improved in recent years and boats can buy licenses for fishing in these waters.

The unilateral extension by these and other countries clearly indicated that an international convention concerning the sea should be held. In 1949 the United Nations had its International Law Commission begin a study on sev-eral marine problems, including the law of the sea. The report, issued in 1956, argued in favor of a 3-mile territorial sea and a contiguous zone out to 12 miles and also proposed that an international conference be held to con-sider the report (International Law Commission Report, 1956).

The next major step in the evolution of the ownership of the sea came from two conferences held in 1958 and 1960 at Geneva. In 1958, represen-tatives from 86 countries met in Geneva to consider the report of the Inter-national Law Commission and to participate in the United Nations Confer-ence on the Law of the Sea. The 1958 conference resulted in four basic con-ventions which were eventually approved by many of the participants:

1. Convention on the Territorial Sea and the Contiguous Zone
2. Convention on the Continental Shelf
3. Convention on the High Seas
4. Convention on Fishing and the Conservation of the Living Resources of the High Seas

Convention on the Territorial Sea and the Contiguous Zone

The first convention established that "the sovereignty of the state extends beyond its land territory and its internal waters to a belt of sea adjacent to its

[3] CEP Doc. no. 2, September 23, 1955 contained in: "Santiago negotiations on fishing conser-vation problems." U.S. Department of State, Public Service Division, 1955, p. 30–32.

coast, described as a territorial sea."[4]   It also established the landward baseline for measuring the breadth of a territorial sea as "the low-water line along the coasts as marked on large-scale charts officially recognized by the coastal state."[4]   However, it did not give any width for the territorial sea. It stated, in Article 6, that "the outer limit of the territorial sea is the line every point of which is at a distance from the nearest point of the baseline equal to the breadth of the territorial sea."[4] Its legal language therefore avoided one of the key issues of the conference. The territorial seas were also defined in a similar manner around islands. Waters on the landward side of the baseline of the territorial sea were defined as forming part of the internal waters of the state. This convention recognized coastal state sovereignty of the air space over the territorial sea as well as of the sea bed and subsoil.

Consideration was also given to what happens when the coasts of two states are opposite or adjacent to each other, in which case neither of the two states is entitled, without agreement between them, to extend its territorial sea beyond the equidistance line which divides the two states.   This provision does not apply in areas where historic title or other circumstances defines the territorial seas between the two states.

The contiguous zone, an area of high seas contiguous to a coastal state's territorial sea, according to the convention "may not extend beyond twelve miles from the baseline from which the breadth of the territorial sea is measured."[4] Within the contiguous zone, the coastal state may exercise control necessary to "prevent infringement of its customs, fiscal, immigration, and sanitary regulations within its territory or territorial sea and punish infringement of the above regulations committed within its territory or territorial sea."[4]   Although the breadth of the territorial sea was not actually defined in this convention, it can be argued that, because of the limit to the contiguous zone, it cannot be greater than 12 miles in width.

Within its internal waters and territorial sea, a coastal state has essentially the same judicial authority as it has in its land territory. It can, if it wants, exclude foreign and alien ships, except for the right of foreign ships to innocent passage within the territorial sea. Passage is considered "innocent" if it does not harm the peace, good order, or security of the coastal state.

In summary, after the Geneva Convention, the countries of the world generally divided their waters into three or four main categories (Fig. 3-1):

1. *Internal waters.* This includes rivers, lakes, and all waters landward of the inner limit of territorial seas (low-water line) as well as estuaries, harbors, and bays between the coast and offshore islands. The coastal state has complete sovereignty over its internal waters, except where these waters had once been territorial or high seas, in which case the right of innocent passage or other historical rights may be allowed.

[4]1958 Convention on the Territorial Sea and the Contiguous Zone—Convention on the Law of the Sea Adopted by United Nations Conference in Geneva, 1958 (U.N. Doc. A/Conf. 13/L. 52).

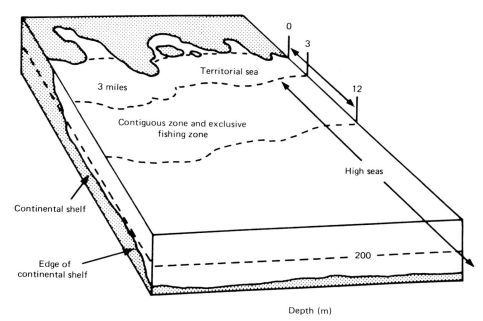

Figure 3-1. General divisions of the ocean after the ratification of the 1958 Geneva Convention.

2. *Territorial Seas.* This is a zone of undefined width, although many states claimed 3 miles at this time. The rights of the coastal state in the territorial sea are similar to that of internal waters except that innocent passage, innocent overflight, and entry under distress conditions are allowed.

3. *Contiguous Zone and exclusive fishing zone.* This zone, according to the Convention on the Territorial Sea and the Contiguous Zone, does not extend beyond 12 miles from where the territorial sea starts. Within this zone, the coastal state may control access for fishing and in this regard may also control fishing research. The contiguous zone is part of the high sea.

4. *High Seas.* The high seas are defined as all waters not included within territorial or internal waters. Within the high seas almost complete freedom exists.

The width of the territorial sea had been extended by some countries to 12 miles, or at least many countries accepted a 9-mile exclusive fishery zone combined with a 3-mile territorial sea. It should be noted that in many instances the 12 miles covered only a portion of a country's continental shelf. As previously mentioned, some countries claimed a 200-mile width for their territorial sea.

The Convention on the Continental Shelf

The Convention on the Continental Shelf defined the shelf as "the seabed and subsoil of the submarine areas adjacent to the coast but outside the area of the territorial sea to a depth of 200 meters, or beyond that limit, to where the superadjacent waters, admits of the exploitation of the natural resources of the said areas; (and) to the seabed and subsoil of similar submarine areas adjacent to the coasts of islands."[5] This clause is one of the more critical ones, in that it says that the coastal state may claim as much of the continental shelf or deeper waters as it is capable of mining. This is commonly referred to as the "exploitability clause."[5] Thus, this convention did not actually define where the continental shelf ends.

The Convention on the Continental Shelf gave the coastal state sovereign rights over the continental shelf "for the purpose of exploring and exploiting natural resources."[5] Natural resources consisted "of the mineral and other non-living resources of the seabed and subsoil together with living organisms belonging to sedentary species, that is to say, organisms which, at the harvestable stage, are immobile on or under the seabed and are unable to move except in constant physical contact with the seabed or the subsoil."[5] The convention also stated that "the coastal State may not impede the laying or maintainence of submarine cables or pipelines on the continental shelf."[5] Furthermore, in Article 5, it stated that "the exploitation of its natural resources must not result in any unjustifiable interference with navigation, fishing or the conservation of the living resources of the sea, nor result in any interference with fundamental oceanographic or other scientific research carried out with the intention of open publication."[5]

Article 5 further said, concerning coastal research, that:

> "the consent of the coastal State shall be obtained in respect of any research concerning the continental shelf and undertaken there. Nevertheless, the coastal State shall not normally withhold its consent if the request is submitted by a qualified institution with a view to purely scientific research into the physical or biological characteristics of the continental shelf, subject to the proviso that the coastal State shall have the right, if it so desires, to participate or to be represented in the research, and that in any event the results shall be published."[5]

This consent requirement concerning freedom of scientific research was to become an obstacle to some research programs; this point is discussed in a later section.

Article 6 concerns the possibility of the continental shelf being adjacent to the territories of two or more States whose coasts are opposite each other. It says

> the boundary of the continental shelf appertaining to such States shall be determined by agreement between them. In the absence of agreement, and unless another boundary line is justified by special circumstances,

---

[5]1958 Convention on the Continental Shelf—Convention on the Law of the Sea Adopted by the United Nations Conference at Geneva, 1958 (U.N. Doc. A/Conf. 13/55).

the boundary is the median line, every point of which is equidistant from the nearest points of the baselines from which the breadth of the territorial sea of each State is measured.[5]

This latter point became important in later years when resources were found in some of the marginal seas lying between several States, such as the mineral deposits of the Red Sea (see Chapt. 5).

Convention on the High Seas

The Convention on the High Seas defined the term "high seas" as "all parts of the sea that are not included in the territorial sea or in the internal waters of a State."[6] Article 2 said that the high seas were

> open to all nations, no State may validly purport to subject any part of them to its sovereignty. Freedom of the high seas is exercised under the conditions laid down by these articles and by the other rules of international law. It comprises, *inter alia* (among others), both for coastal and noncoastal States:
>
> 1. Freedom of navigation;
> 2. Freedom of fishing;
> 3. Freedom to lay submarine cables and pipelines;
> 4. Freedom to fly over the high seas.
>
> These freedoms, and others which are recognized by the general principles of international law, shall be exercised by all States with reasonable regard to the interests of other States in their exercise of the freedom of the high seas.[6]

This convention also had two articles concerning pollution. One, Article 24, said "Every State shall draw up regulations to prevent pollution of the seas by the discharge of oil from ships or pipelines or resulting from the exploitation and exploration of the seabed and its subsoil, taking account of existing treaty provisions on the subject."[6] Article 25 said that

> "Every State shall take measures to prevent pollution of the seas from the dumping of radio-active waste, taking into account any standards and regulations which may be formulated by the competent international organizations. All States shall cooperate with the competent international organizations in taking measures for the prevention of pollution of the seas or air space above, resulting from any activities with radio-active materials or other harmful agents."[5]

Convention on Fishing and Conservation of Living Resources of the High Seas

The convention on Fishing and Conservation of Living Resources of the High Seas concerned mainly the biologic resources of the ocean and said that

[6]1958 Convention of the High Seas–Convention on the Law of the Sea Adopted by the United Nations Conference at Geneva, 1958 (U.N. Doc. A/Conf. 13/L. 53).

All States have the right for their nationals to engage in fishing on the high seas, subject (a) to their treaty obligations, (b) to the interests and rights of coastal States as provided for in this Convention, and (c) to the provisions contained in the following articles concerning conservation of the living resources of the high seas. All States have the duty to adopt, or to cooperate with other States in adopting such measures for their respective nationals as may be necessary for the conservation of the living resources of the high seas.[7]

The 1958 Conference was followed by another in 1960. The results of these conferences were in later years to cause controversy on several important aspects, in particular: (1) the lack of definition of the width of the territorial sea; (2) the 200-m depth outer limit of the continental shelf; and (3) the exploitability of marine mineral resources. The failure to delineate the width of the territorial sea permitted some individual coastal States to make their own determination of its width. In some instances, this resulted in the claiming of territorial seas on the order of hundreds of miles in width. States that did not ratify the conventions, of course, were not bound by them. The second controversy was that the definition of the continental shelf usually had little to do with the geologic or oceanographic characteristics of the shelf. The use of the 200-m depth contour is essentially arbitrary; in some instances the edge of the continental shelf is much shallower and in some instances, much deeper. A depth definition was not acceptable for countries with no shelves or very narrow shelves, such as Peru and Chile, or to some countries that have shelves that extend for literally hundreds of miles but are more than 200 m deep.

The third controversy resulting from the conferences was the ambiguity of the exploitability component of the definition of the continental shelf: It stated that the shelf extended to "a depth of 200 meters or, beyond that limit, to where the depth of the superadjacent waters admits of the exploitation of the natural resources of the said areas."[8] The exploitability concept left the ownership of the sea floor open to all kinds of interpretations, including that an entire ocean could be divided among the countries bordering it (Fig. 3-2), or, for example, that the United States or any other country which had the capability could mine its shelf, continental slope, rise, ocean basin, and ridge, and perhaps right up on to the other side of the ocean to another country's shelf. The United States has already leased areas for oil exploration and exploitation in depths greater than 200 m off California and at over 160 km off the coast into the Gulf of Mexico. Based on the exploitability clause, therefore, one could easily extend a claim into the deep sea. This point was apparently not as important to the delegates in 1958 and 1960 as it

---

[7]Convention on Fishing and Conservation of Living Resources of the High Seas—Convention on the Law of the Sea Adopted by the United Nations Conference at Geneva, 1958 (U.N. Doc. A/Conf. 13/L. 54).

[8]See footnote 4.

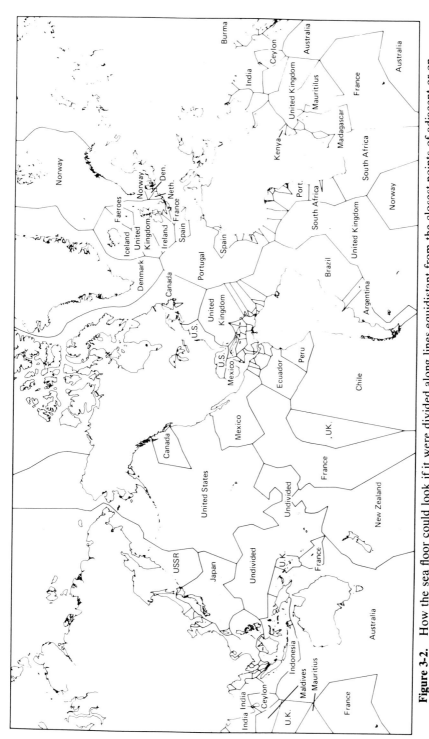

Figure 3-2. How the sea floor could look if it were divided along lines equidistant from the closest points of adjacent or opposite coastal States. This could be a possible, but improbable, division based on the exploitability concept. (Adapted from Christy and Herfindahl 1967. Base chart is H.O. 1262F)

is today, where the promise of exotic mineral and biologic resources in the deep ocean has caused many countries to show considerable concern about exploitation. Chapter 5, on the resources of the ocean, discusses some of these problems and shows that in some instances, the resources of the deep sea are generally not as valuable as anticipated.

In one respect, these conferences were very important in that it was probably the last time that the high seas would be treated as an area of unqualified freedom. Later debates have seen countries either claim ownership of parts of the deep sea or try to establish international regimes over the high seas. Following the conference, over 100 nations indicated that they had some jurisdiction over the minerals in the areas adjacent to their coasts. This confirmation has taken many forms, including individual and unilateral declarations, domestic legislation, treaties, offshore concessions, or actual ratification of the 1958 Convention on the Continental Shelf. As it has turned out, over 40 countries (including most of the major maritime powers) have actually ratified this convention. Each of the four conventions became effective 30 days after the 22nd State had agreed to it; the Convention on the Continental Shelf reached this point in June 1964.

One irony to these conferences was that the United States, perhaps realizing the Pandora's Box it had opened by the Truman Proclamations, argued very strongly on the behalf of a 3-mile limit to territorial seas (6 miles for the 1960 conference) but was not successful in convincing other countries to define a definite width to the territorial sea. In the 1960 conference, a proposal for a 6-mile wide territorial sea failed by one vote.

Events Leading to the Law of the Sea Conferences in the 1970s

The period between 1960 and the early 1970s saw considerable change in how nations viewed the ocean. The developing countries continued to press their claims for further ownership and participation in the supposed riches of the sea and by 1973 as many as 17 states had claimed territorial seas of more than 12-naut. mi. width. It was also during this period of time that marine technology made several advances, especially in offshore drilling, that made the "exploitability" clause more important. Scientific research and fishing operations were also involved in conflicts during this interval. On an optimistic note, it was suggested during this period that the resources of the ocean should belong to all countries or be "the common heritage of mankind." These pressures on the use and the territorial claims of the ocean eventually resulted in a series of Conferences on the Law of the Sea that started in late 1973 and are still in progress. The following discussion concerns the events leading up to these conferences and some of the results and conflicts.

The years immediately following the 1958 Geneva Conference were essentially a transition period until the conventions were ratified in the early 1960s. Even so, not all States participated or abided by the spirit of the agreement. In addition, advances in marine technology made parts of the sea bed beyond 200-m depth available for mineral exploration. Among the technologic advances were offshore drilling for oil in water up to 1000 m deep and the developing technology to mine the manganese nodules on the deep-sea floor (see Chapt. 5). In addition, expansion and mechanization of distant-water fishing fleets allowed them to stay at sea for months. Therefore, the legal vacuum created by the Geneva Conference on the ownership of resources beyond 200-m depth started to become critical.

During these years the United States was also reevaluating its marine policy with regard to the high seas and its mineral resources. President Johnson, in a July 1966 speech for the commissioning of a new oceanographic research vessel, said that

> under no circumstances must we ever allow the prospects of rich harvest and mineral wealth to create a new form of colonial competition among the maritime nations. We must be careful to avoid a race to grab and hold the lands under the high seas. We must ensure that the deep seas and the ocean bottoms are, and remain the legacy of all human beings.[9]

Although the United States was seriously concerned about the extension of claims into the oceans, it was the representative of the small country of Malta who provided a dramatic impetus toward a reconsideration of policy toward the ocean. In August of 1967, the Permanent Mission of Malta proposed that the 22nd Session[10] of the General Assembly of the United Nations add to its agenda an item concerning the development of a treaty that would reserve the sea bed and ocean floor beyond the limits of national jurisdiction and their resources "in the interest of mankind." This proposal and a subsequent speech by the Maltese Ambassador, Dr. Arvid Pardo, noted the rapid technologic advances by developed countries which could cause the ocean floor and its resources to be subjected to national appropriation. He wished the ocean floor to be reserved for peaceful uses and that its resources become "the common heritage of mankind." The financial resources derived from this part of the ocean outside national jurisdiction could be used to further the development of less developed countries. This proposal, which was fairly controversial, received essentially unanimous support and led to a resolution that created in December 1967 an *Ad Hoc* Committee to Study the Peaceful Uses of the Sea Bed and Ocean Floor Beyond the Limits of National Jurisdiction. The *ad hoc* committee was ini-

[9]The President's remarks at the Commissioning of the New Research Ship, The *Oceanographer*, July 13, 1966, 2 Wkly. Comp. Pres. Docs. 930, 931 (1966).
[10]U.N. Doc A/6695, dated August 18, 1967.

tially composed of 35 nations[11] and was to prepare a report for the 23rd Session (1968) of the General Assembly covering:

1. A survey of past and present activities of the United Nations, its specialized agencies, and other intergovernmental bodies regarding international agreements concerning the sea bed and ocean floor
2. An account of the scientific, technical, economic, legal, and other aspects of the sea bed and ocean floor
3. An indication concerning practical methods of developing international cooperation in the exploration, conservation, and use of the sea bed and ocean floor

The United Nations *ad hoc* sea bed committee met several times in 1968 and 1969 without reaching any major agreement. In December of 1969, after considerable debate and opposition by the United States and the Soviet Union, the United Nations adopted (65 to 12 with 30 abstentions) a resolution to convene

> a conference on the law of the sea to review the regimes of the high seas, the continental shelf, the territorial sea and contiguous zone, fishing and conservation of the living resources of the high seas, particularly in order to arrive at a clear, precise, and internationally accepted definition of the sea bed and ocean floor which lies beyond national jurisdiction, in the light of the international regime to be established for that area.[12]

The sea bed committee was expanded to include a total of 86 members (later the Peoples Republic of China and four other countries joined, for a total of 91 members). The United States was not too enthusiastic at this time about a new convention, because it felt that the 1958 Geneva Conventions were sufficient concerning United States ownership of resources on the continental margin, and especially because the exploitability clause allowed the United States to claim what it had the technologic capacity to reach.

At about the same time the General Assembly also passed a resolution, commonly called the Moratorium Resolution, which stated that "nations are bound to refrain from all activities of exploration of the resources of the area of the sea bed and ocean floor, and the subsoil thereof, beyond the limits of national jurisdiction."[13] Although the Moratorium Resolution was not legally binding, the United States voted against it, feeling that it would retard deep-sea technologic development.

Another important event concerning the ocean floor occurred in February

---

[11]The following member States were appointed to the *ad hoc* committee: Argentina, Australia, Austria, Belgium, Brazil, Bulgaria, Canada, Ceylon, Chile, Czechoslovakia, Ecuador, El Salvador, France, Iceland, India, Italy, Japan, Kenya, Liberia, Libya, Malta, Norway, Pakistan, Peru, Poland, Romania, Senegal, Somalia, Thailand, USSR, Tanzania, United Arab Republic, United Kingdom, United States, and Yugoslavia. All except Austria and Czechoslovakia are coastal States.
[12]United Nations Document A/2574A (1969).
[13]United Nations Document A/2574D (1969).

1971 when the United States, USSR, England, and 60 other nations signed the Treaty on the Prohibition of the Emplacement of Nuclear Weapons and other Weapons of Mass Destruction on the Sea Bed and Ocean Floor and Subsoil Thereof. This treaty prohibited the placing of nuclear weapons and associated structures beyond the maximum contiguous zone as defined in the 1958 Geneva Convention, or in other words outside of 12 naut. mi. from where a country begins its territorial sea. It does not prevent the emplacement of such weapons within the territorial sea or contiguous zone (out to a total of 12 miles).

The *ad hoc* committee continued to meet in 1970, although again it was usually unable to agree on major recommendations or proposals. The chairman of the committee, H.S. Amerasinghe from Sri Lanka (Ceylon), was finally able to get a committee consensus on a Declaration of Principles Governing the Sea Bed and the Ocean Floor, and the Subsoil Thereof, beyond the Limits of National Jurisdiction which was unanimously adopted by the General Assembly in December of 1970. The main point of this declaration[14] was the reaffirmation that the resources beyond the limits of national jurisdiction are the common heritage of humankind. It also said that all resource activities in this area beyond national jurisdiction were to be governed by an international regime. On the question of scientific research in the area beyond national jurisdiction, it recommended that states "promote international cooperation in scientific research for peaceful purposes"[14] and to develop "measures to strengthen research capabilities of developing countries, including the participation of their nationals in research programs."[14] The declaration did not comment on where the boundary between national jurisdiction and the international zone should be. Perhaps even more important, the General Assembly at the same time approved a resolution to have an international conference on the law of the sea. The vote on the resolution, which was proposed by the United States, was passed by the General Assembly by 109 to 7, with 6 abstentions; those abstaining or voting against the resolution were mostly the Soviet block countries.[15]

The plan was to have the Conference start in 1973 (a December 1973 meeting only elected officers and developed an agenda but did not establish procedural rules) and to have the *ad hoc* Sea Bed Committee serve as the preparatory committee. The committee was to be responsible for the agenda, location, and date of the meeting. In March of 1971, at its 45th meeting, the Committee on the Peaceful Uses of the Sea Bed and the Ocean Floor Beyond the Limits of National Jurisdiction (or Sea Bed Committee) divided itself into three subcommittees. Subcommittee I was concerned

---

[14]United Nations Document A/RES/2479 (1971).

[15]Voting against were Bulgaria, Byelorussian Soviet Socialist Republic, Czechoslovakia, Hungary, Poland, Ukrainian Soviet Socialist Republic, and Union of Soviet Socialist Republic. Abstaining were Burma, Cuba, Romania, Saudi Arabia, and Venezuela.

with the sea bed area beyond national jurisdiction and what would be the powers and functions of any international regime that administered this area. Subcommittee II was to prepare a list of issues for the conference to consider, including width of the territorial sea, regime of the high seas, international straits, and fisheries. Subcommittee III was concerned with protection of the marine environment, i.e., pollution and scientific research. These three subcommittees were to consider principles and arguments and present draft treaties to the convention. Many countries submitted draft articles on the different subjects and considerable data and political exchange followed, which ideally was to have led to compromises that would be satisfactory to all or most of the participants. This generally did not turn out to be the case; in fact, not one single draft treaty on a major item was agreed on before the 1974 meeting of the conference. Actually, the considerable ranges of interests among the participants in the conference and their desires indicated that agreement would be difficult. The 1974 conference had twice the number of participants than the 1958 and 1960 Geneva Conferences. More important is the fact that many of these countries were developing nations who wanted a change in the old freedom of the seas concept. This is because these developing countries felt that the old systems allowed the few developed countries to dominate the ocean and its resources.

Basic Issues of the Law of the Sea Conferences of the 1970s

The basic points facing the Law of the Sea Conferences were as follows:

1. The width of the territorial sea and adjacent economic zone and degree of coastal state control.
2. Ownership of resources of the water, sea bed, and subsoil. This item is of two main parts: the resources within the area under the coastal state's jurisdiction (and therefore the question of the extent of this jurisdiction); and the resources in the area beyond national jurisdiction or in the deep or high seas.
3. Right of navigation and overflight through what are now international straits but which could be included within expanded territorial seas or an economic resource zone.
4. Management of living resources in the ocean, especially those species that migrate and those in areas where many countries have traditionally fished for them but that now may come under a single country's jurisdiction.
5. Protection and reduction of pollution in the ocean.
6. Freedom of scientific research in the ocean.

7. A regime for control or management of the high seas.

The United States, early in the negotiations, indicated its willingness to accept a 12-mile territorial sea with the condition of an international guarantee of free and unimpeded passage through international straits. This is a critical point to maritime countries because there are 116 straits that have a high-sea passageway under a 3-mile territorial sea (because they are wider than 6 miles) that would become entirely territorial seas if there were a change from a 3- to a 12-mile territorial sea (see discussion in Chapt. 4 and Table 4-11, page 100). These include some very important straits, such as Gibraltar, Malacca, and Bab el Mandeb (southern end of the Red Sea). One of the major concerns of countries having large navies is that if these straits become territorial seas and there is no guarantee of unimpeded passage, navies would have only the right of innocent passage, which would require submarines to travel through them on the surface and not allow military planes to fly over them. This could severely limit the United States' military position around the world; its acceptance of a 12-mile territorial sea is very strongly locked to free passage.

The strait issue is also very important to oil-importing countries, such as Japan, whose trade may be seriously affected by unreasonable transit regulations by coastal states through whose straits their ships have to pass. Likewise, the coastal State may want to establish pollution requirements within their newly acquired waters as well as exercise some control over military operations.

The new width of the territorial sea and the economic resource zone will undoubtedly be 12 and 200 naut. mi., respectively (Fig. 3-3). Other past possibilities included a territorial sea out to a depth of 200 m and an economic resource zone out to a depth of 3000 m which, in some instances, corre-

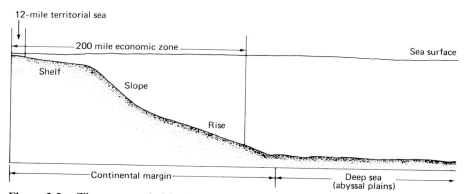

Figure 3-3. The most probable division of the sea floor arising from the Law of the Sea Conference. Compare this figure with Fig. 3-1, which shows the division of the sea floor after the 1958 Geneva Convention. See later portions of this chapter for a discussion of some proposed legal definitions of the shelf.

Table 3-1. Major Net Gains of Some Individual Countries Depending on Possible Divisions of the Sea Floor[a]

Country	Shelf Area to 200-m Depth		Shelf Area to 200 Naut. mi.	
	Sq. Naut. Mi.	km²	Sq. Naut. Mi.	km²
Canada	846 500	2 837 300	1 307 600	4 367 000
Indonesia	809 600	2 704 000	1 577 300	5 268 200
Australia	661 600	2 209 700	2 043 300	6 824 600
United States	545 400	1 821 600	2 222 000	7 421 400
USSR	364 300	1 216 700	1 309 500	4 373 700

[a]Adapted from Albers and Meyers (1974).

sponds to the edge of the continental margin (shelf, slope, and rise). The net gain to individual States would vary considerably, with Canada having the greatest net gain with a 200-m zone, Australia the most in a zone to 3000 m, and, ironically, the United States the winner in the 200-mile sweepstakes (Table 3-1). The total amount of land gained under the different possible scenarios is considerable: for 200 m, 8 031 000 sq. naut. mi.; for 3000 m, 24 039 000 sq. naut. mi.; and for 200 naut. mi., 37 745 000 sq. naut. mi. The larger area gained by the 200-naut. mi. width is certainly one of the reasons that this value was chosen. These numbers represent a considerable portion of the ocean or of the land (Table 3-2). It should be mentioned that each of these depths or distances has no significant oceanographic or geologic meaning associated with it. Many countries have continental shelves shallower than 200 m; several are considerably deeper. Likewise, a distance criteria of 200 miles would in some instances, such as for Argentina and the USSR, not reach over all parts of their shelves. Obviously some countries will not be satisfied by the 200-mile exclusive economic zone. Those countries that feel they would be disadvantaged by such a solution are looking for some type of compensation, such as income from resources found on shelves of those countries that have "better" shelves. The exact legal definition of the shelf is also a source of difficulty (discussed further on page 62).

Table 3-2. Percentage of Total Ocean or Total Land That Would Result from Various Divisions of the Ocean (Not Including Antarctica)[a]

	200 m	3000 m	200 naut. mi.
Percent of total ocean	7.6	22.8	35.8
Percent of total land	18.5	55.4	87.0

[a]Adapted from United States Department of State (1972).

There are other problems, such as variations and seasonal changes in the shoreline, which is the starting point for boundary measurements, and how the seaward boundaries are to be marked or indicated in the open sea, where may ships are without sophisticated navigational devices. It has been suggested by some (especially Hedberg 1976b) that several boundaries be established; for example, one for navigational purposes (for the ocean surface), one for fishing rights (for the water), and one for mineral rights (the sea floor and below). This relatively reasonable suggestion could ultimately be considered in final negotiations.

One of the more interesting aspects of the 200-mile boundary is how it would actually be defined. One cannot put posts or markers out in the open ocean, so the boundary will be hard to establish physically. This problem is compounded by the difficulty of positioning a marine vessel: The 200-mile distance is too far from land for radar, and LORAN is not usually available to all ships and, even so, it has an accuracy of plus or minus a mile; satellite navigation systems are not much better. Using water depths as markers would not work because many charts are not that accurate. In addition, the charts are based on depths that do not take into account possible sound velocity changes (influenced by temperature and salinity of the water) and so the measured depths can change because of changes in the conditions of the overlying water. Besides the limitation problem for the ship that may be intruding into the zone, there are similar problems for the country that may be defending or monitoring its own zone.

It is beyond the scope of this book to discuss the position of each country (even if it were known) concerning the law of the sea. However, some basic considerations relate to the views of several interest groups. Perhaps most important is that many of the developing countries lack important knowledge of their offshore environment and therefore know little of their resource potential. In many instances, this is because of lack of marine scientific expertise within the country. Likewise, some states feel that total freedom of the seas, beyond the area of national jurisdiction, offers the most advantages to those countries that are sufficiently technologically advanced so that they can use and exploit the resources in this area. Based on these two points, therefore, it is logical that less developed countries wish to have an extensive or exclusive economic resource zone, over which they have considerable control, as well as to have a say in the development of any resources from outside the resource zone. It should also be noted that developing countries generally put a much higher priority on achieving economic growth by using marine resources than on being concerned about environmental quality and pollution of the sea. Many developing countries view pollution as a problem of the developed countries, in that they have caused much of it by their industrial development and in that they are the ones that can afford to deal with this problem.

The participants in the conference can be divided into various interest groups; a single country can belong to several. Each interest group would favor what most beneficially affected them. For example, coastal States would favor a broad economic resource zone to gain control over offshore oil and gas as well as fishing rights. Many of these States would not favor sharing any income with other States. Landlocked States (Table 3-3), which have no access to the ocean, would tend to favor a narrow economic resource zone with as much of the resources as possible to be shared under the common heritage principle. Shelflocked States (because of geography these States have only a small offshore area that cannot be extended

Table 3-3. Landlocked and Shelflocked Countries[a]

Landlocked (30)	Shelflocked[b] (19)
Afghanistan	Bahrain
Andorra	Belgium
Austria	Cambodia
Bhutan	Denmark
Bolivia	Finland
Botswana	Germany (East)
Burundi	Germany (West)
Central African Republic	Iraq
Chad	Jordan
Czechoslovakia	Kuwait
Hungary	Monaco
Laos	Netherlands
Lesotho	Poland
Liechtenstein	Qatar
Luxembourg	Singapore
Malawi	Sweden
Mali	Togo
Mongolia	United Arab Emirates
Nepal	Vietnam (North)
Niger	
Paraguay	
Rhodesia	
Rwanda	
San Marino	
Swaziland	
Switzerland	
Uganda	
Upper Volta	
Vastican City	
Zambia	

[a]Adapted from Alexander (1973).
[b]Those with offshore areas whose depth does not exceed 200 m.

seaward) would favor only a modest extension of national jurisdiction. Technologically advanced countries would prefer a regime where they could use their skills to obtain the offshore resources without concern for a changing international position about their ownership of the mining area. Fishing States, especially those that fish off other countries' coasts, would favor limited extension of national jurisdiction or some mechanisms that would allow them to continue their present fishing operations. States that border on strategic international straits want control over ship's passage through these bodies of water. States with large navies would prefer to see present international straits remain open. This is not a complete list of the different possible interest groups, but it can be seen how one state could fall into several groups that have conflicting interests.

The general consensus concerning the Law of the Sea Conference is that at best many years will pass before an international agreement may possibly be reached and signed. Recent events occurring in the 1977 and 1978 conference, such as attempts to establish very restrictive conditions for manganese nodule mining (see Chapt. 5 for details) outside of the 200-mile exclusive economic zone, have suggested to many that a comprehensive treaty may not really be possible or, if it is, it may not be acceptable to some developed countries. In late 1977 and early 1978 the United States Government held hearings to consider withdrawing from the conference but decided not to at that time.

The principle part of any agreement will be that a large portion of the ocean, out to 200 naut. mi. from the coast, is added in some exclusive manner to the adjacent coastal state. In addition, other territory may be added if the Irish or some other formula (see later sections of this chapter) is accepted. Ironically, this division of the ocean will make Dr. Pardo's original suggestion of the common heritage of the ocean be but a hollow dream, because most of the oil, gas, and biologic resources that are in the ocean occur within this 200-mile economic zone (see Chapt. 5). The remaining part of the ocean is, with the exception of manganese nodules, essentially devoid of known significant resources. The manganese nodules can be mined for their copper, manganese, nickel, and cobalt content; however, the ratio of these elements within the nodules is badly out of phase with the ratio of the use of these metals by industry (see Table 5-8). This, in effect, limits the amount of nodules that can be mined in a given year. As discussed in more detail in Chapter 5, if 10% of the value of the copper, manganese, cobalt, and nickel mined in the world (estimated at $10 billion for 1980) could come from nodules and this amount could be taxed at 10%, only $100 million would be available for "the common heritage of all mankind." This works out to about $715,000 for each of the 140 or so countries belonging to the United Nations. Only if the countries of the world were to share the resources within the 200-mile zone off their coasts could the common heritage concept have the meaning for which Dr. Pardo had hoped.

The meeting of the Law of the Sea Conference held in Caracas in 1974 ended with very little to show other than plans to meet the following year in Geneva. Many of the delegates felt that the views between different interests had been narrowed and that the next step was government to government meetings and negotiations. Others considered this meeting to be a failure and there was considerable concern that the lack of any progress would encourage countries to make unilateral extensions of their national jurisdiction.

The 1975 meeting of the Law of the Sea Conference in Geneva likewise did not produce a treaty or even come close to one, although the group agreed to meet again in New York in 1976. A draft treaty (diplomats refer to it as the "informal single negotiating text (A/Conf. 62/WP. 8)") was released after the 1975 meeting, but it had not been voted on, and it was not even clear whether any major points had majority support. One major problem of the conference was the desire of developing countries to have an international sea-bed authority to control resource development and open ocean research. It was generally agreed among the delegates of the 1975 meeting that territorial seas would be extended to 12 miles and that some sort of economic resource zone out to 200 miles would be established.

After this and later conferences, many individual countries took unilateral action to extend their claims in the sea (Table 3-4). A significant move in this direction was made in the United States by legislation filed by Representative Studds of Massachusetts and Senator Magnuson of Washington. Their bill extended United States' fishing jurisdiction beyond its past 12-mile limit to 200 miles off the United States coast. The U.S. Department of State opposed the bill because it felt the bill would encourage other countries to make similar moves that might endanger the conference. In spite of these arguments, the Fishery Conservation and Management Act (P.L. 94-265) was signed and went into effect on 1 March 1977 (see Chapt. 5 for details concerning the legislation). Several other countries also established similar 200-mile zones; for example, Mexico in mid-1976, Canada on 1 January 1977, and the European Economic Community countries on 1 January 1978.

After the 1974 and 1975 conferences, both the United States and the Soviet Union, as well as several other major maritime powers, still tied their acceptance of the 200-mile exclusive economic zone to guarantees of free and unimpeded passage through international straits by merchant and military ships, as well as overflight by aircraft. However, many of the major marine nations made compromises in some of their original positions in an attempt to resolve other questions and problems.

One last point should be mentioned: Even though the majority of countries at the Law of the Sea Conference might favor a particular position, without the concent of the major maritime powers it would be hard to establish any sort of international legal regime. However, the major maritime

Table 3-4. Changes in Territorial Sea Claims in the Period 1973–1977[a]

Breadth (nautical miles)	Number of Countries		
	1973	1975	1977 (September)
3	25	30	26
4	4	4	4
6	11	13	8
10	1	1	1
12	54	57	60
15	1	0	1
18	1	1	0
20	0	0	1
30	3	4	3
50	1	3	4
100	0	1	2
130	1	1	0
150	0	1	2
200	10	9	14
Modified archipelago	0	3	3
No legislation	6	1	1

[a]Data from U.S. Department of State reports.

powers generally know they cannot make ocean law without the support of the less developed countries. Both sides must cooperate or little will happen.

The following two meetings, in 1976 (the new draft document was called the Revised Single Negotiating Text, A/Conf. 62/W.P. 9) and 1977 (new draft document was called the Informal Composite Negotiating Text A/Conf. 62/W.P. 10), both in New York, resulted in even further divisions between the developed and less developed countries, and further disagreement on an acceptable treaty. When the 1977 Informal Composite Negotiating Text was released the United States Law of the Sea Ambassador, Elliott Richardson, said that the terms relating to ocean mining of manganese nodules were fundamentally unacceptable. He also noted serious, unresolved problems, such as lack of assurance of access to the deep-sea bed, requirements for technology transfer, unreasonable financial burdens and production limits placed on ocean mining, and open-ended powers to be conferred upon a sea-bed authority.

The 1978 meeting held in April in Geneva was an especially difficult one. It started with a controversy concerning the presidency of the conference. In previous sessions Mr. Amerasinghe of Sri Lanka was president, and many delegations would have preferred him to remain in this capacity.

However, a change in the government of Sri Lanka had occurred and the new government wanted Mr. Amerasinghe removed from their delegation.

The first 2 weeks of the meeting were devoted to a debate of whether Mr. Amerasinghe should stay or leave (a position favored by many Latin American countries). Finally it was concluded that he should stay for that meeting and a "resumed" session that followed at the end of the summer in New York.

Two major areas of disagreement continued to dominate the meeting. These were how to reconcile the concern of those states who felt that a 200-mile exclusive economic zone (EEZ) was not in their best interests and what should be the character of an International Sea-Bed Authority and what control it should have over the mining of the manganese nodules in the deep sea (see also Chapt. 5 for comments on this subject). A smaller issue, but one of considerable concern to marine scientists, was the freedom of science question (discussed in a later part of this chapter).

The problem concerning the EEZ is that some countries have margins wider than 200 naut. mi., whereas others are landlocked or have aspects of their offshore region that makes a 200-mile zone unappealing. It has been suggested that the "disadvantaged" states should receive some income derived from the resources of margins off states that are more advantaged. The idea obviously is not favored by many of the "advantaged" states. Several formulas have been proposed for defining the outer limits of the continental margin (continental shelf, slope, and rise) when it exists beyond 200 miles. These include:

1. The Arab formula, which says the edge of the continental margin should be the same as the EEZ or 200 naut. mi.

2. The Soviet formula, which would permit states to extend their jurisdiction of the margin if it extended beyond 200 miles to an additional 100 miles if it could be justified by geologic or geomorphic evidence

3. The Irish formula, which recommended that jurisdiction could extend to 60 miles beyond the foot of the continental slope or, alternatively, to where the thickness of sediments is at least 1% of the distance between that point and the foot of the slope

A revised Informal Composite Negotiating Text (A/Conf. 62/W.P. 10/Rev. 1) developed at the 1979 meeting in Geneva defines the continental shelf (the term used by the convention that includes the continental shelf, slope, and rise) in terms of the Irish Formula, but prohibits it from extending more than 350 naut. mi from the coast or 100 naut. mi from the 2,500 m isobath, whichever is greater.

Perhaps even more difficult to resolve is the question of control over the deep-sea bed. Several states support the concept of having an authority with jurisdiction over sea-bed resources as well as over such items as research. The principle sea-bed resource is manganese nodules (see pages 137–155), which contain valuable amounts of manganese, copper, nickel, and cobalt. Many of the developing countries of the world prefer having an authority control these resources (and sharing the profits) than seeing them mined by technologically advanced nations (in particular the United States). The United States has proposed a parallel type of mining system, whereby half of the mining sites would be developed by private companies or individual states and half would be under the control of an international sea-bed authority. The United States Congress has been considering, for several years, various legislation to provide an interim mechanism for sea-bed mining by United States companies until a consensus is reached in the Law of the Sea negotiations (see page 154). This possible action is viewed with considerable concern by many delegations, especially those from lesser developed countries (often called the Group of 77, although their number may be higher). This group, itself, has many diverse opinions on almost all subjects. The United States argues that all countries presently have the rights to explore and exploit deep-sea resources because it is one of the freedoms of the high seas. The proposed United States legislation would not take precedence over a law of the sea treaty in which the United States participated.

At present, considerable differences exist between the delegates on the authority issue. As of this writing, the next session (the eighth) will be in New York in the summer of 1979. Whether this will be the last one or just another is not clear. Consensus must be reached on the above issues as well as others in such areas as fishing rights (especially as they relate to migratory species and to permitting other countries to fish in one's EEZ—see pages 174–176 in Chapt. 5), navigation, pollution, boundaries (see page 55, Chapt. 4), and dispute settlements.

In all fairness, it should be stated that these meetings are probably among the most complicated and difficult ever held, and that they really have turned out to be more of a confrontation of development and economics than a look at peaceful and harmonious ways of using the ocean. One unfortunate aspect was that the question of freedom of scientific research quickly got damaged in the negotiations. This is discussed further in the next section.

The Freedom of Scientific Research Issue

The degree of freedom for scientific research is an especially important aspect in the Law of the Sea negotiations, especially because the area where research seems most likely to be limited—the coastal region of the oceans—is in many aspects the most important scientific zone of the ocean. The concept of freedom for scientific research in the ocean has generally been treated as an accepted fact by marine scientists essentially since the study of the oceans began. Nowhere, however, is this "freedom" stated in any official international document. In the past, the main exception to freedom of scientific research was in a foreign country's inland waters and territorial sea, where they have complete sovereignty but where permission for scientific research can be given. The 3-mile width of the territorial sea was really not challenged until the Truman Proclamation of 1945, when the United States extended its jurisdiction over sea-floor resources out to a depth of 200 m. This proclamation was followed by ones from several Latin American countries, who extended their jurisdiction to as much as 200 miles off their coasts and, in some instances, claimed the right to regulate scientific research within these newly claimed areas. The 1958 and 1960 Geneva Conferences, as previously stated, did not reach any agreement on the width of the territorial sea. The following years saw additional countries extending their claims further seaward (see Table 3-4).

The 1958 Convention on the Continental Shelf states that the

consent of the coastal State shall be obtained in respect of any research concerning the continental shelf and undertaken there. Nevertheless, the coastal State shall not normally withhold its consent if the request is submitted by a qualified institution with a view to purely scientific research into the physical or biological characteristics of the continental shelf, subject to the proviso that the coastal State shall have the right, if it so desires, to participate or to be represented in the research, and that in any event the results shall be published.[16]

During the 1950s and 1960s, as countries were extending their claims, the effect on scientific research was not very evident. When countries had restrictions these could usually be alleviated or removed by contacting scientific colleagues in the country, inviting them to come on the cruise, and sometimes in this way gaining permission to make the study. Another reason, at that time, for not worrying about occasional restrictions was that, because so many areas of the ocean had not been studied, if permission was not given in one area the scientific program could easily be modified to study other, equally interesting localities. In the last decade, however, there has been a noticeable increase in the restrictions by countries on scientific research. The problems of getting permission for scientific research can have diplomatic aspects, for example, the U.S. Department of State, up to the late 1970s, did not recognize territorial seas greater than 3 miles in width or contiguous zones wider than 12 miles. Therefore, if a United States scientist wanted to work in the waters of Peru, which claims a territorial sea of 200 naut. mi., the U.S. Department of State generally would not request permission unless the scientist agreed to work within 3 miles of the Peruvian coast. In this case the U.S. Department of State could request permission to work in Peru's territorial sea without acknowledging their 200-mile claim.

In 1972 Dr. Conrad Cheek of the U.S. Department of State made a sur-

[16]Article 5, Part 8: 1958 Convention on the Continental Shelf (U.N. Doc. A/Conf. 13/L. 55).

vey of the United States' position on scientific research for the then pending 1973 Law of the Sea Conference (Cheek 1972). Concerning clearances, 28 respondents indicated rejections of clearance requests and there were 22 abandonments of requests because of delays, etc. Cheek determined that 357 requests had been granted which, on the basis of his numbers, suggested that only one out of eight requests had failed. Many of the clearances granted had some restrictions attached to them. For example, of the 357 clearances granted, there were 275 coastal state scientists whose participation was invited before the clearance was requested, 80 scientists whose uninvited participation was required by the coastal state, and 35 nonparticipating observers whose presence was mandatory. There were relatively few requests for substantial changes in research plans, deletion of projects, or changes in cruise tracks or station patterns, except in the east Asian region, especially the USSR. It was also noted that of the 60 then pending requests, 52 were being negotiated through the U.S. Department of State. This was interpreted to mean that the coastal states had begun to press their claims on controlling research more exclusively through diplomatic channels.

Cheek also inquired into the attitudes of marine scientists toward the possible restrictions that coastal states might impose upon their research programs. His survey showed that most United States scientists did not object to the coastal state having some participants aboard and even to giving them some training. There was substantial objection, however, to giving the coastal states control over the data, samples, and publication rights and to allowing them to make major modifications to the research program.

When the position of a coastal state is considered, there are understandable reasons that it may have concern about foreign scientists working within its waters. Among these is confusion over the difference between scientific research and exploration; therefore, some countries feel that research could have a detrimental effect on their control over their own resources. Many do not have the proper scientific expertise to evaluate the scientific findings. Some states feel that marine research can be a form of espionage and so may jeopardize a country's national security. In general, these views have been appreciated by marine scientists, and it is realized that oceanography, in general, would benefit if developing countries could acquire a competent level of marine science capability. This would give them an understanding of basic marine research and allow meaningful participation in the the actual marine research program, as well as improving opportunities for joint research programs. Unfortunately, there are at present no major funding programs available for the training of foreign marine scientists.

If research in the EEZ were to be reduced it would be critical, because for oceanographers to understand the important processes of the ocean it is imperative that they be able to consider the boundary conditions where land and water meet, both along the coastline and on the continental shelf. This area is the source of sediments for the deep sea, where waves and currents end, and where most upwelling of ocean water occurs, and it is the area of input of most pollutants and other chemicals into the ocean. Indeed, many ocean phenomena are global in scale and are not affected by or do not respect political boundaries. If oceanographers could only study the deep sea or that area outside the 200-mile limit, oceanography could become a relatively sterile and esoteric science.

As we shall see in later sections, the question of freedom of scientific research is not always an all or nothing question. In some instances the restrictions put on research are reasonable and can be met with little jeopardy or delay to the program. In other instances, which unfortunately are increasing in number, the restrictions are such that they effectively make the research nearly impossible.

The area of the ocean that would be affected by increased regulations on scientific research is considerable. Figure 3-4 shows what is excluded if research is restricted in an area of 200-m water depth or less, as has been suggested for a legal definition of the continental shelf; this area is equal to 7.6% of the ocean. If, instead, as seems almost guaranteed, the entire

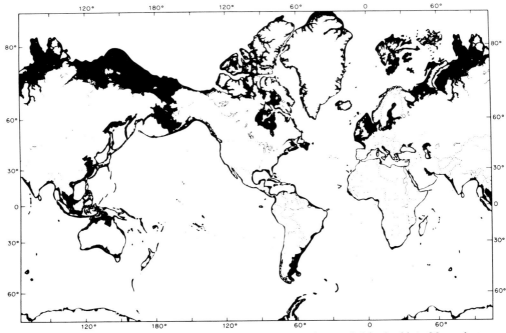

Figure 3-4.   Chart of areas within the 200-m contour interval (blacked in). Note that mercator projection distorts polar regions and that Australia is over three times the size of Greenland.

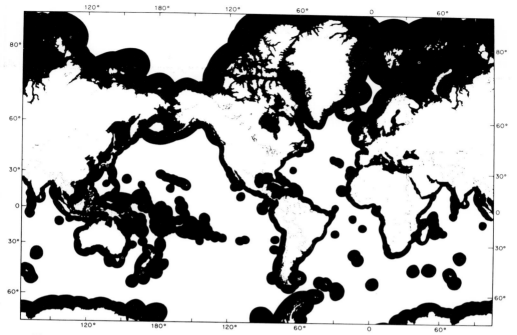

Figure 3-5.   Chart of area (blacked in) that would fall within a 200-naut. mi. economic resource zone.

200-mile economic resource zone is subject to strong and restrictive regulations, then about 35.8% of the ocean, or about 37.7 million nautical square miles, could be eliminated (Fig. 3-5). (If Antarctica were included, the amount would be even greater.) It has even been suggested that research be regulated in the remaining portions of the ocean, such as that included in the Irish formula (see page 62).

When the Sea Bed Committee, in 1971, divided itself into three subcommittees, Subcommittee III was established to deal with the preservation of the marine environment including, *inter alia*, scientific research and prevention of pollution. In later years several marine scientists from many countries have met with this subcommittee and other delegates to the Law of the Sea Conference, both formally and informally, to discuss the mutual advantages of marine science research. In one instance, a research vessel was even brought to the United Nations to demonstrate its use in marine science (Fig. 3-6). On another occasion, delegates were taken on a short 1-day cruise. Dr. John Knauss, Provost of the Graduate School of Oceanography of the University of Rhode Island, made a statement to the Third Preparatory Meeting of the Law of the Sea Conference in August 1972 in which he tried very clearly to show the importance of science to society and said

Figure 3-6.   Visit of Research Vessel *Knorr* (of the Woods Hole Oceanographic Institution) to the United Nations to demonstrate its use to the delegates of the Law of the Sea Convention. (Photograph courtesy of Woods Hole Oceanographic Institution)

> all the technological and economic advances of modern times have been based on the findings of science. A strong case can be made for the thesis that if it were not for the advancements in technology generated by science, there would probably be no need for a new Law of the Sea Conference.[17]

He also addressed the question of basic research and applied or economic research and said

> the distinction between the primary objectives of science and the importance of science to society should be noted. For example, because sci-

[17]Statement by Professor John A. Knauss before Subcommittee III of the Committee on the Peaceful Uses of the Sea Bed and the Ocean Floor Beyond the Limits of National Jurisdiction, 3 August 1972.

ence has led to a wider use of ocean resources, some see a direct relationship between scientific research and resource exploration. There is indeed a relationship but it is not a direct one.[17]

Dr. Knauss also gave several examples of how knowledge obtained from basic research has led to a better understanding of the ocean and has been beneficial to humankind.

To give the reader a better feeling for the debate on scientific research, I mention below some points made by delegates from different countries at the 1973 meeting of Subcommittee III.

Doubts were expressed as to the usefulness of the distinction between fundamental or pure research and applied research or research aimed at commercial exploration. It was argued that what might appear basic and fundamental research in the eyes of one scientist would be research aimed at the exploration of mineral resources to another.

One pointed out that certain national interests relating to security and commercial matters were involved in scientific research.

The opinion was expressed that it was possible to identify pure scientific research investigations with noncommercial and nonindustrial aims. This was in contrast to another view that the concept of pure science was theoretical and a fallacy in the light of international political and social–economic realities.

The expression of freedom of scientific research was not to be interpreted as one of the freedoms of the high sea and should be replaced by the term "promotion and development of scientific research."

Another view held that freedom to conduct scientific research was one of the universally recognized freedoms of the high seas and represented the common principle of customary international law.

Another view was that whereas there was freedom of scientific research and that this freedom should be protected, it should also be subjected to appropriate regulations so that it would take into account the rights and interests of other states and conform with the basic provisions established to protect the marine environment.

Another said the coastal state had the right to control marine scientific research in areas under its jurisdiction and to insure the protection of its vital interests in this regard as well as the duty to promote such research and act as the custodian of the international community's interests in the development of scientific knowledge concerning the marine environment as a whole.

For the marine scientist the Law of the Sea Conference will decide several especially pertinent questions, including:

1. What will be the limits of the territorial sea?
2. If an economic resource zone is established, how wide will it be and what will be the conditions or regulations for scientific research within this zone?

3. What, if any, controls will be put on research in general, and will any regulations be put on work in the high seas?

It would be premature to attempt to predict the eventual result, if there ever will be one, of this debate on the freedom of marine scientific research. However, it is clear that the previous general lack of regulations on marine research will not be continued in the future; in fact, it appears that the control being discussed at the conferences will be sufficient to strongly hinder marine science in the economic zone and perhaps elsewhere. The restrictions could have some small compensatory advantages in that they would encourage oceanographers to develop research programs with underdeveloped countries. However, this can just as easily happen without restrictions. In any case, it probably would be naive to assume that a country might change its position and allow uncontrolled access to its waters because of a single opportunity for a research program. However, it is also clear that those countries that put severe restrictions on basic scientific research will probably only be hurting themselves in the long run.

In spite of all the good intentions and discussions by marine scientists for scientific research, it was never given much support by the delegates from other countries at the conferences. The Informal Composite Negotiating Text resulting from the 1977 meeting especially treated scientific research rather poorly, listing numerous restrictions and leaving many statements so open ended as to make almost any interpretation possible by a coastal state. An example was Article 247, Part (a), which says that coastal states may withhold their consent for marine scientific research in their exclusive economic zone if the project "is of direct significance for the exploration and exploitation of natural resources, whether living or non-living."[18] This statement could be used to cover most any form of basic marine scientific research. The revised Informal Composite Negotiating Text (A/Conf. 62/W.P. 10/Rev. 1) resulting from the 1979 Geneva meeting has only minor changes in the articles concerning marine science.

As Dr. Handler, President of the U.S. National Academy of Sciences, once said in a speech to Subcommittee III:

> To be sure, regulations would not mean the end of oceanography. Research will continue, but, the more it is subject to controls, the greater the danger that it will become second class research. Progress will be slower, research will be more expensive and new avenues of inquiry will not be pursued. By directing science toward short-term goals, we may lose the unique scholar driven only by his own curiosity, who has been responsible for so many of the great scientific discoveries of the past. I submit that mankind cannot afford such a loss.[19]

[18]Informal Composite Negotiating Text, Third United Nations Conference on the Law of the Sea, New York Session, 23 May–15 July 1977, (A/Conf. 62/W.P. 10).
[19]Statement by Dr. Philip Handler to Subcommittee III of the United Nations Committee on the Peaceful Uses of the Sea Bed on the Ocean Floor, 29 March 1973.

References

Alberts, J.P., M.D. Carter, A.L. Clark, A.B. Coury, and S.P. Schweinfurth, 1973, *Summary Petroleum and Selected Mineral Statistics for 120 Countries, Including Offshore Area.* U.S. Geol. Survey Prof. Paper 817. Washington, D.C.: U.S. Government Printing Office, p. 125.

*Alexander, L.M., 1973, Indices of national interest in the oceans: *Ocean Devel. Internatl. Law Jour.,* v. 1, p. 21–49.

Anderson, E.V., 1974, The law of the sea battle—An overview: *Marine Technol. Soc. Jour.,* v. 8, n. 6, p. 4–14.

Brown, S. and L.L. Fabian, 1974, National interests vs. international needs: *Marine Technol. Soc. Jour.,* v. 8, n. 6, p. 29–39.

Burke, W.T., 1970, *Marine Science Research and International Law:* Occasional Paper No. 8, Law of the Sea Institute, University of Rhode Island, Kingston, R.I., 25 p.

Burke, W.T., 1975, *Scientific Research Articles in the Law of the Sea Informal Single Negotiating Text.* Occasional Paper No. 25, Law of the Sea Institute, University of Rhode Island, Kingston, R.I., 16 p.

Cadwalader, G., 1973, Freedom of science in the ocean: *Science,* v. 182, p. 15–20.

*Cheek, C.H., 1973, Law of the sea: Effects of varying coastal states having controls on marine research: *Ocean Devel. Internatl. Law Jour.,* v. 1, n. 2, p. 209–219.

*Christy, F.T. Jr., and H. Herfindahl, 1967, *A Hypothetical Division of the Sea Floor.* Chart prepared for the Law of the Sea Institute, University of Rhode Island, Narragansett, R.I.

Duomani, G.A., 1971, Exploiting the resources of the seabed: *In: Science, Technology and American Diplomacy.* Report prepared for the U.S. House of Representatives, Subcommittee on National Security Policy and Scientific Developments of the Committee on Foreign Affairs. Washington, D.C.: U.S. Government Printing Office, 152 p.

Federal Ocean Program, 1973, Washington, D.C.: U.S. Government Printing Office.

Fye, Paul M., 1972, Ocean Policy and Scientific Freedom: Columbus O'Donnell Iselin Memorial Lecture, Washington, D.C.: Marine Technology Society, 11 September 1972.

Griffin, W.L., 1969, Law of ocean space: *In:* Firth, F.E., Ed., *Encyclopedia of Marine Resources.* New York: Van Nostrand Reinhold, p. 347–352.

*Handler, P., 1973, Statement to Subcommittee III on the Peaceful Uses of the Sea Bed and the Ocean Floor; United States Mission to the United Nations, New York, 29 March 1973.

Hargrove, J.L., 1973, New concepts in the law of the sea: *Ocean Devel. Internatl. Law Jour.,* v. 1, n. 1, p. 5–12.

Hedberg, H.D., 1973, The national–international jurisdictional boundary on the ocean floor: *Ocean Management,* v. 1, p. 83–118.

Hedberg, H.D., 1976a, Ocean Boundaries for the Law of the Sea: *Marine Technol. Soc. Jour.,* v. 10, n. 5, p. 6–11.

---

*References indicated by an asterisk are cited in the text; others are of a general nature.

*Hedberg, H.D., 1976b, Ocean boundaries and petroleum resources: *Science,* v. 191, p. 1008–1018.

Hollick, A.L., 1974, The clash of U.S. interests: How U.S. policy evolved: *Marine Technol. Soc. Jour.,* v. 8, n. 6, p. 15–28.

Hurd, B., and B. Passero, 1974, *Oceans of the World: The Last Frontier.* An annotated introductory bibliography on the law of the sea: Rept. No. MIT Sea Grant 74-17, 10 p.

International Law Commission Report, 1956, United Nations General Assembly, Official Record, 11th Session, Supp. n. 9 (A/3159).

Knauss, J.A., 1973, Developing the freedom of scientific research issue of the Law of the Sea Conference: *Ocean Devel. Internatl. Law Jour.,* v. 1, n. 1, p. 93–120.

Knauss, J.A., 1974, Marine science and the 1974 Law of the Sea Conference—Science faces a difficult future in changing Law of the Sea: *Science,* v. 194, p. 1335–1341.

Knight, H.G., 1973, United States ocean policy—Perspective 1974: *Notre Dame Lawyer,* v. 49, n. 2, p. 241–275.

Krueger, R.B., 1971, An evaluation of United States oceans policy: McGill Law Jour., v. 17, n. 4, p. 604–698.

Llana, C.B., J.K. Gamble Jr., and C. Quinn, 1975, *Law of the Sea: A Bibliography of the Periodical Literature of the 1970's.* Law of the Sea Institute, Kingston, R.I., Special Publication No. 4.

Miles, E., and J.K. Gamble, Eds., 1977, *Law of the Sea: Conference Outcomes and Problems of Implementation.* Cambridge, Mass. Ballinger: 425 p.

Oxman, B.H., 1978, The Third United Nations Conference on the Law of the Sea: The 1977 New York Session: *Amer. Jour. Internatl. Law,* v. 72, p. 57–83.

Qasim, S.Z., 1973, Development of marine science capabilities in different regions of the world: *In* Bologna Conference Report, held by the Johns Hopkins University, 15–19 October 1973.

Shepard, F.P., 1973, *Submarine Geology,* 3rd ed. New York: Harper & Row, 517 p.

United States Department of State, 1972, *International Boundary Study: Limits in the Seas—Theoretical areal allocations of seabed to coastal states.* The Geographer, Office of the Geographer, Bureau of Intelligence and Research, No. 42, 35 p.

United States, 1969, *Marine Science Affairs:* The Second Report of the President to the Congress on Marine Resources and Engineering Development, 228 p.

Wooster, W.S., 1973a, Scientific aspects of maritime sovereignty claims: *Ocean Devel. Internl. Law Jour.,* v. 1, n. 1, p. 13–20.

Wooster, W.S. (Ed.), 1973b, *Freedom of Oceanic Research.* New York: Crane, Russak, and Co.

Chapter 4

Marine Shipping

Introduction

Shipping, among all the uses of the sea, is probably the most indispensible to society. For example, most of the world's foreign trade travels on ships, a method of transportation that is about 40 times cheaper than air transport. Shipping of oil to the United States, Japan, and countries in Europe from the Middle East, Venezuela, and other areas is of critical importance for the very survival of these energy-dependent economies (Fig. 4-1, Table 4-1). Shipping is the largest item in terms of total ocean dollar expenditures, with close to $100 billion being spent each year by the countries of the world to build, maintain, and repair ships.

Surprisingly, some of the major trading countries do not have very large merchant fleets (Table 4-2), although the numbers are a little deceptive because many companies or individuals maintain a foreign registry for their ships (Liberia and Panama are the most popular). For example, between 35% and 40% of Liberian registered ships are United States' owned. Among the reasons for foreign registry are sometimes less stringent safety requirements and certain legal and financial advantages. International law recognizes the ownership of a ship by its flag, not by the nationality of the owner, its port of call, or even where taxes on it are paid. In some instances the registry and ownership of ships are purposely made vague to avoid responsibility in case of accidents or negligence. *Torrey Canyon* (Fig. 6-5), according to Mostert (1974), was owned by the Barracuda Tanker Corporation, which was a subsidiary of Union Oil Corporation. Union had subleased it to British Petroleum Trading Ltd., a subsidiary of British Petroleum Company.

Figure 4-1.   The primary flow of oil. (Courtesy of the American Petroleum Institute)

Table 4-1. Interregional Oil Imports and Exports: 1975–1977 (In Millions of Tons)[a]

	Imports			Exports		
	1975	1976	1977	1975	1976	1977
United States	300	365	437	11	12	16
Canada	42	39	36	40	30	27
Caribbean	67	78	76	144	161	147
Other western hemisphere	49	46	43	16	14	17
Western Europe	626	682	661	12	17	24
Middle East	13	12	7	918	1033	1028
North Africa	9	7	4	119	147	158
West Africa	2	3	7	98	111	118
East and South Africa, South Asia	44	45	47	2	1	1
Southeast Asia	78	88	86	72	84	91
Japan	246	262	271	—	—	—
Australasia	17	17	17	2	3	3
USSR, Eastern Europe, and China	15	19	23	74	93	101
Destination not known[b]	—	43	14	—	—	—
Total	1508	1706	1729	1508	1706	1729
Of which seaborne	1468	1675	1694	1468	1675	1694

[a]Source: British Petroleum Trading Ltd. From: Organization for Economic Cooperation and Development (1978), p. 47

[b]Including increased quantities in transit, transit losses, minor movements not otherwise shown, military use, etc.

The ship itself was built in the United States, rebuilt in Japan, registered in Liberia, and insured in London, and it had an Italian crew.

The World Shipping Fleet

The number and size of the ships engaged in worldwide commerce has increased considerably since World War II. The increase has been variable, depending on trade and technology. In the past few years emphasis had been on building large supertankers; more recently it has turned toward liquefied natural gas (LNG) carriers (Fig. 4-2 and Table 4-3). Supertankers are the biggest ships afloat (Table 4-4 and Fig. 4-3 and 4-4). Their immense size, however, presents harbor, navigation, and steering problems, which are

Table 4-2. Major Merchant Fleets of the World as of 1975[a,b]

Country of Registry	Total Number	Total Gross Tons[c] (1000s)	Total dwt Tons[d] (1000s)	Number Tankers	Number Freighters	Number Bulk Carriers
Liberia	2 358	60 006	112 086	945	506	841
USSR	2 358	13 533	17 278	455	1 413	150
Japan	2 143	35 994	60 167	520	951	542
Greece	1 838	22 339	36 665	344	909	479
United Kingdom	1 609	32 153	52 980	467	650	332
Panama	1 341	11 539	18 513	222	870	183
Norway	1 028	25 095	43 306	337	313	320
United States	922	12 503	17 637	278	539	19
(Private)	(583)	(9 820)	(14 446)	(245)	(313)	(19)
(Government)	(339)	(2 683)	(3 191)	(33)	(226)	(—)
West Germany	668	8 494	13 616	84	460	77
Total All Countries	22 449	306 366	503 348	5 121	11 449	4 075

[a]Source: Maritime Administration Reports (1976, 1977), United States Department of Commerce.

[b]Includes all ocean-going ships over 1000 gross tons; excludes ships used in inland waters and military owned merchant ships. Tonnage is in thousands.

[c]Gross tons is a volume measurement; each gross ton equals 100 ft^3 of enclosed space.

[d]Deadweight tonnage, or dwt, is the number of long tons (2240 lb) of fuel cargo, etc., that a ship can carry.

Figure 4-2. The *Norman Lady*, a liquified natural gas transport ship, which with her five spherical tanks carries the energy equivalent of 540 000 barrels of oil. The gas is carried at a temperature of $-260°$F. The vessel has been in operation since 1973. (Photograph courtesy of International Nickel Company)

Table 4-3. World Fleet by Type of Vessel and Regional Distribution, Mid-1977[a]

Type of Vessel	Total Tonnage; (million G.R.T.)	Percent Increase/ Decrease, Mid-1976/ Mid-1977	Percentage of Type Registered In:			
			OECD Countries[b]	Flag of Convenience Countries[c]	USSR/ Eastern Europe[d]	Rest of the World
Oil tankers	174.12	+3.5	53.8	34.9	3.3	8.0
Combination carriers	26.09	+4.3	60.7	31.9	1.1	6.3
Ore and bulk carriers	74.83	+12.2	59.1	27.8	4.2	8.9
General cargo ships	77.09	+4.1	45.3	20.3	13.4	21.0
Container ships	7.54	+12.8	86.9	8.0	1.4	3.7
Passenger vessels and ferries	7.09	−5.5	73.8	6.7	10.1	9.4
Liquefied gas carriers	4.41	+30.6	61.4	32.0	1.0	5.6
Chemical tankers	1.76	+37.8	75.8	14.0	0.2	10.0
Fishing and fish-carrying vessels	12.16	+2.6	30.0	1.4	59.0	9.6
Miscellaneous[e]	8.59	+16.7	60.2	11.6	13.6	14.6
All types	393.68	+5.8	54.2	27.8	7.3	10.7

[a]Source: Lloyd's Register of Shipping. From: Organization for Economic Cooperation and Development (1978). p. 62.
[b]Including Great Lakes fleets, and the United States Reserve fleet (almost entirely general cargo ships).
[c]Cyprus, Lebanon, Liberia, Panama, Singapore, Bahamas, and Oman.
[d]Albania, Bulgaria, Czechoslovakia, Germany (Democratic Republic), Hungary, Poland, Romania, and USSR.
[e]Including lighter carriers, vehicle carriers, livestock carriers, research vessels, supply ships, tugs, dredgers, cable ships, icebreakers, and other miscellaneous ships.

Table 4-4. Size of Largest Tankers, Bulk Carriers and Passenger Ships (as of 28 July 1975)[a]

Name and Country of Registry	Deadweight Tonnage	Length (ft)	Breadth (ft)
Oil Tankers			
Nissei Maru, Japan	484 337	1242	203
Globtik London, Britain	483 939	1243	203
Globtik Tokyo, Britain	483 664	1243	203
Ioannis Colocotronis, Greece	386 612	1213	210
Hemland, Sweden	372 201	1193	208
Nisseki Maru, Japan	366 813	1138	179
Bulk Ore, Bulk Oil, and Ore Oil Carriers			
Sevaland, Sweden	282 450	1109	179
Dececanyon, Liberia	271 235	1113	180
Jose Bonifacio, Brazil	270 358	1106	179
Tarfala, Sweden	265 000	1099	170
Mary R. Koch, Liberia	265 000	1099	170
Passenger Ships			
Queen Elizabeth II, Britain	66 852[b]	963	105
France, France	66 348	1035	110
Raffaello, Italy	45 933	904	101
Michelangelo, Italy	45 911	904	101
Canberra, Britain	44 807	818	102

[a]Source: Lloyd's Register of Shipping 1974.
[b]In gross tons.

discussed in other sections of this chapter. The large tankers are generally called VLCC (very large crude carriers) if they are between 200 000 and 400 000 tons and ULCC (ultra large crude carriers) if they are more than 400 000 tons.[1]

At the end of 1973, the world's tanker fleet stood at well over 200 million deadweight tons and about 40 million tons (Table 4-5) were added in 1974; this one year's increase was essentially equivalent in size to the entire world's tanker fleet in 1957. In recent years, there has been a dramatic reversal in the need and construction of new supertankers (Table 4-5). Just before the 1973 Israeli–Arab war there were not enough available, and it was during this period that plans for many new VLCC and ULCC tankers were developed. The Arab oil embargo, fuel conservation programs, and the worldwide recession, however, lowered the need for new tankers, and by

[1]The size of a ship is generally designated either by its length or by its tonnage. Passenger ships generally use gross tonnage, which is their enclosed cubic capacity, with one gross ton being equivalent to an enclosed space of 100 ft³. If the capacity is registered with a country the capacity may be indicted as gross registered tonnage (G.R.T.). For a passenger ship a large portion of the tonnage may be used as cabins and not necessarily for cargo. Tankers generally are described in terms of deadweight tons (dwt). Deadweight tonnage is the weight of fuel, cargo, ballast, ship stores, etc. (but not the weight of the ship itself), that can be carried.

CROSS SECTION OF A VERY LARGE CRUDE CARRIER

Length . . . . 1141 ft
Beam moulded . . . . 170 ft
Draft on summer freeboard . . . . 65½ ft
Deadweight . . . . 253 000 tons
Sea speed . . . . 16 knots

1. Universal chock
2. Vents
3. Mooring winch
4. Fairleader
5. Rudder stock
6. Rudder
7. Mooring bitts
8. Four bladed propeller
9. Companion ladder
10. Swimming pool
11. Stern light
12. Provisions hoist
13. Engineroom upper level
14. Boiler flat
15. Auxiliaries
16. Propeller shafting
17. Thrustbearing
18. Gear case

19. HP and LP turbines
20. Condenser
21. Main boiler
22. Cooling water overboard discharge
23. Diesel generator
24. Uptake
25. Settling tanks
26. Ballast tanks
27. Pumproom
28. Lifeboat davit
29. Bridge wing
30. Radar and signal mast
31. Wheelhouse
32. Pumproom vent
a. Walkway
b. Ramp
c. Breakwater
33. Separator
34. Portside longitudinal bulkhead
35. Flume stabilization openings in longitudinal bulkheads
36. Mainlines to pumproom
37. Center keelson
38. Starboard longitudinal bulkhead

39. Deck longitudinals
40. Tank brackets
41. Walkway
42. Cable tray
43. Foam monitors on platform
44. Deck lights
45. Reinforced webframe in center tank
46. Webframes in wingtanks
47. Deck girder in wingtank
48. Bulkhead with stringers and vertical webs
49. Operating panel for hose derricks
50. Mast with lights and hose derricks
51. Accomodation ladders stowed midships
52. Hoist for accomodation ladder
53. Center deck girder
54. Platforms and ladders in cargotanks
55. Side keelson in wingtanks
56. Butterworth holes
57. Hose saddle
58. Guard for hydraulic lines
59. Hose though
60. Fixed tank-washing machine
61. Crossover

62. Tank hatch
63. Panama chock
64. Longitudinal frames
65. Swash bulkhead
66. Side longitudinals
67. Center tank
68. Foremast with deck lights
69. Anchor windlasses
70. Companion
71. Hatch
72. Tank vent
73. Chain roller with devil's claw
74. Forepeak bulkhead
75. Forepeak
76. Warning light for bulbous bow
77. Anchor light
78. Light waterline

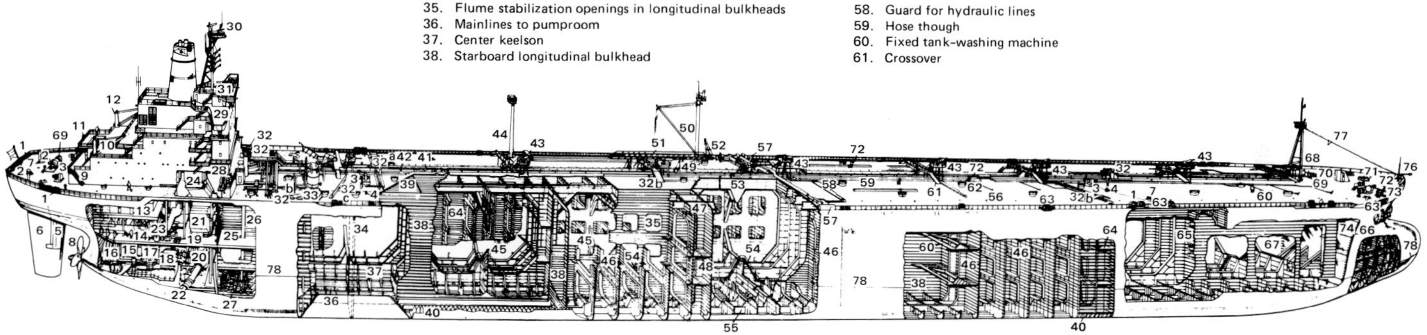

Figure 4-3. Cross section of a very large crude carrier (VLCC). Courtesy of the American Petroleum Institute—an EXXON photograph)

Figure 4-4. Rudder and screw of a VLCC (very large crude carrier). Compare their sizes to the man in the foreground. (Photograph courtesy of the American Petroleum Institute—an EXXON photograph)

1975 there were more than enough ships available. For example, in December 1974, 21 tankers, three of them supertankers, were laid up and by August 1975, as many as 442 ships, including numerous supertankers, were taken out of operation. By July 1977 the tonnage of laid-up ships was 35 million deadweight; in addition, 35 million deadweight tons were being absorbed by slow steaming and 15 million deadweight tons surplus to miscellaneous inefficiencies and combined carriers. Slow steaming (going at low speeds) is a technique used to increase usage of ships as well as reduce port

Table 4-5. Actual and Estimated Tanker Deliveries[a]

Year	Tonnage (dwt)
1972	20.6
1973	28.2
1974	41.8
1975	44.4
1976	50.3
1977	26.4
1978	9.2
1979	2.4

[a]Data from the *Oil and Gas Journal*, 28 June 1976, p. 124.

congestion (over 90 days wait in some Middle Eastern ports in 1978). The maiden voyage of some supertankers has been a trip to a mothballing area. By 1978 the worldwide tanker fleet exceeded 380 million deadweight tons.

The glut of supertankers will probably get worse before it gets better, because some ships are still on order, with deliveries to spread out over $2\frac{1}{2}$ years. The present demand now seems to have turned toward smaller ships that can use the Suez Canal, work the North Sea area, or carry oil from the Alaskan pipeline terminal at Valdez to the relatively shallow-water ports of the west coast of the United States.

One result of the Arab oil embargo and the realization of the worldwide energy shortage has been the development of liquified natural gas (LNG) systems. In many oil-rich areas, such as the Persian Gulf, natural gas used to be treated as a waste product and flared or burned off at the well site. Natural gas has the unique property that its volume can be reduced about 600 times when it is liquified (liquification occurs below a temperature of $-260°F$ at atmospheric pressure). If kept at or below this temperature, therefore, 3 billion cubic feet of gas can be stored or transported in a 900 000-barrel tank.

Because of the above characteristics of natural gas and its abundance and need, there has been an increase in the building of LNG tankers (Fig. 4-2). By mid-1975 the international LNG carrier fleet consisted of 65 vessels in actual service or under construction, but only 20 of these had actually been delivered and only 14 were in regular service.

Another important development in shipping is the use of container ships. In this method of shipping, cargoes are put into large, usually rectangular, containers that can be easily loaded on and off a vessel. This system has the advantage of rapid transfer to and from the port area, ease of handling, less use of port space, less possibility of loss or pilferage, and the need for smaller crews because of automation. A more advanced system is the so-called lighter aboard ships (LASH), which carry loaded barges that can be moved around from ship to ship or ship to port.

Nuclear-powered ships are a type of vessel that has not realized its full potential, except for nuclear-powered submarines (see Chapt. 7). Among the advantages for nuclear ships are that they do not have to refuel (and therefore spend less time in port) and probably are environmentally safer (with respect to oil pollution) than conventional ships, although this point is challenged by many. There are presently five nonmilitary nuclear vessels. One, the Japanese *Mutsu*, had its sea tests delayed for almost 2 years because of concern about its safety, and when these tests were finally begun in 1974, the nuclear reactor was found to leak. The vessel drifted for days with typhoons raging at sea and political storms raging on land; fishermen were especially concerned about potential pollution to their scallop beds. Four years later, in 1978, the *Mutsu* left, under normal power, from its home port (Mutsu City) to the southern part of Japan. It will spend 3 years in that area undergoing repairs to its reactor.

The United States freighter *Savannah* had similar concern expressed for its pollution potential but has been inoperative since 1970 because of high operating costs. The *Otto Hahn*, a West German ore carrier, and the *Lenin* and, most recently, the *Arktika*, both Soviet icebreakers, have experienced less controversy and apparently more successful careers. In August 1977 the *Arktika* made history by being the first surface vessel to reach the North Pole. To get there the vessel had to break through ice that was sometimes as much as 12 feet thick.

It should be noted, perhaps sadly, that the day of the ocean-going passenger ship is essentially over. The crossing of the major oceans by passenger ship has been almost eliminated by the speed, convenience, and economy of air travel, although it is still occasionally possible to travel as a passenger on a cargo ship or to take local trips aboard cruise ships.

Other types of vessels, such as recreational boats, ferryboats, and such innovative ships as hydrofoils or hovercraft, because of their design and use, do not spend much of their time on the open ocean and so are not considered here. Many of these types of vessels, however, require special facilities, such as marinas, and constitute a major industry that, combined with the great number of boat owners, accounts for a large use of the sea.

The United States Shipping Fleet

To better understand how a nation's fleet can develop I shall present a brief discussion of the evolution and change of the United States' merchant fleet.

The United States' ocean-going merchant fleet has changed considerably in size over the past 40 years. Prior to World War II it had a fairly large, but old fleet, because many of the ships had been built for World War I. During World War II vast numbers of ships were sunk, but the United States had an intense ongoing shipbuilding program that resulted in a quadrupling of its fleet between 1941 and 1954. It was possible during World War II to produce a liberty ship complete with launching in 14 days. By the end of the war the United States' fleet was larger than that of the rest of the world combined.

Because most other countries had their fleets destroyed during the war, a large amount of worldwide shipbuilding was initiated in the late 1940s and early 1950s. The United States helped this effort by foreign aid programs and by selling, rather cheaply, many United States' merchant vessels. Shipbuilding in the United States did not keep pace with the selling of used ships or with replacing old and sometimes obsolete vessels (Table 4-6). By 1975 the United States' fleet stood eighth in total number of ships, seventh in gross tonnage, eighth in deadweight tonnage, and seventh in number of tankers. (Again, it should be stated that many United States owned ships were flying flags of convenience. Therefore, figures for national ownership are often somewhat misleading.) The Soviet shipping fleet in 1950, for comparison, had only about 430 vessels and ranked 21st in size in the world, but by 1975 they were very near the top. More important, the present Soviet fleet is relatively young (almost all fewer than 20 years old), whereas much of the United States' fleet is over 20 years old. In addition, the Soviets carry more than half their own foreign trade, and that percentage is increasing, whereas the United States carries only about 5% of its foreign trade, although the percentage is also increasing. As an oceanographer it is very common for me to see several Soviet tankers or freighters in foreign ports, whereas it is rare to see more than one United States' ship in the same port.

The relatively small number (and capacity) of United States' flag tankers has contributed to the United States balance of payments problem, because more than 95% of its biggest import, petroleum, is carried on foreign ships (many may actually be United States owned vessels flying flags of convenience). In mid-1977 the United States President was supporting legislation that would eventually require 9.5% of all imported oil to be carried in United States ships. Such regulation was supported by maritime interests, such as United States shipbuilders' and merchant seamen's organizations. The proposed legislation would have slowly raised the United States' percentage by only 0.5% per year in the first few years to 9.5% by 1982. Such actions, of course, can lead to retaliation by foreign governments and legislation against United States ships. There is precedence, however, for such United States action. France, for example, requires that 66% of its imports be carried in French ships. The U.S. House of Representatives did not share the President's enthusiasm for cargo-preference legislation and in October of 1977 defeated the proposed legislation. It should be pointed out that the United States is the largest owner and operator of container ships, which carry a large value of the total (nonoil) world's trade.

On the other hand, the United States shipping industry is one of the most highly subsidized industries. It has been estimated by Jantscher (1975) that since World War II United States maritime industries have received over

Table 4-6.　Age Distribution of World Fleet as at Mid-1977 (ships of 100 G.R.T. and over)[a]

Country	Under 5 Years	5–10 Years	10–15 Years	15–20 Years	20–25 Years	25–30 Years	30 Years and Over	Total Tonnage 1000 G.R.T.
Australia	35	34	15	10	4	0	2	1 374.2
Belgium	45	21	23	8	1	1	1	1 595.5
Canada[b]	10	17	28	14	7	8	16	2 822.9
Denmark	57	20	15	5	2	0	1	5 331.2
Finland	41	26	16	12	5	2	1	2 262.1
France	50	32	11	6	1	0	0	11 613.9
Germany (F.R.)	42	37	13	5	2	0	1	9 592.3
Greece	18	24	20	20	12	4	2	29 517.1
Italy	35	22	18	12	7	3	3	11 111.2
Japan	37	43	15	4	1	0	0	40 035.9
Netherlands	25	27	20	21	6	0	1	5 290.4
Norway	47	34	14	4	1	0	0	27 801.5
Portugal	39	21	12	16	4	5	3	1 281.4
Spain	50	29	10	6	2	1	2	7 186.1
Sweden	60	27	8	3	1	0	1	7 429.4
Turkey	36	19	9	14	7	3	11	1 288.3
United Kingdom	42	36	11	7	2	1	1	31 646.4
United States[c]	24	15	8	12	7	3	31	15 299.7
OECD Countries[d]	37	31	14	9	4	1	4	212 479.5
Algeria	77	15	31	3	1	1	0	1 056.0
Argentina	17	6	13	23	9	13	19	1 677.2
Bermuda	48	22	21	4	2	2	1	1 751.5
Brazil	48	28	6	12	1	1	4	3 330.0

Bulgaria	30	19	26	7	10	4	4	964.2
China (P.R.)	16	12	32	21	9	3	7	4 245.4
Cyprus	1	3	7	28	38	16	7	2 787.9
Germany (D.R.)	10	16	34	27	8	5	0	1 486.8
India	44	17	22	8	6	1	2	5 482.2
Indonesia	15	8	10	28	18	12	9	1 163.2
Iran	44	34	3	3	4	0	2	1 002.1
Iraq	87	11	0	1	0	0	1	1 135.2
Kuwait	60	28	6	3	1	1	1	1 831.2
Liberia	45	28	15	8	4	0	0	79 983.0
Panama	30	16	12	18	13	7	4	19 458.4
Philippines	27	11	12	23	13	5	9	1 146.5
Poland	43	25	19	10	2	1	0	3 447.5
Romania	60	15	19	4	1	0	1	1 218.2
Saudi Arabia	34	22	23	5	8	5	3	1 018.7
Singapore	55	9	6	12	11	6	1	6 791.4
South Korea	37	30	12	6	9	3	3	2 494.7
Taiwan	8	49	13	12	7	3	8	1 558.7
U.S.S.R.	22	24	30	14	7	1	2	21 438.3
Yugoslavia	27	23	22	17	7	3	1	2 284.5
Rest of the World	30	17	16	18	10	4	5	12 446.2
World Total	38	27	15	10	5	2	3	393 678.4
Of which:								
Tankers[e]	48	26	13	8	3	1	1	174 124.4
Dry bulk carriers	37	34	18	6	2	1	2	100 921.6
Others	23	24	16	15	11	5	6	118 632.4

[a]Source: Lloyd's Register of Shipping. From: Organization for Economic Cooperation and Development (1978). p. 140.
[b]Including the Great Lakes fleet.
[c]Including the Great Lakes fleet and the reserve fleet.
[d]Excluding Austria, Iceland, Ireland, New Zealand, and Switzerland, for which figures are not given by Lloyds.
[e]Including combination carriers.

$10 billion in subsidies. These subsidies can come as tax benefits and shelters, construction and operating subsidies, and guarantees of government and private business.

Within recent years the condition of the United States maritime industry has improved. This resulted from several factors, one of which was the devaluations of the dollar in 1971 and 1973 that made United States shipping prices more competitive with foreign prices. Other factors are record farm exports and oil imports, improvement in labor relationships, and the Merchant Marine Act of 1970, which provided subsidies for the construction of new ships. United States shipbuilding industries have been pioneers in building some types of sophisticated ships, such as oil drilling (Fig. 5-7 and 5-9) and container ships. In the larger volume and profit areas of building cargo ships, however, American shipyards in the past just were not able to construct ships as cheaply as foreign yards. Government subsidies, although available and substantial, did not apply to the building of large tankers until the passage of the Merchant Marine Act of 1970. This act also revised the system for shipbuilding and ship operating subsidies, to encourage increased efficiency. At first the subsidy was to pay up to 45% of the cost difference in building the ship in an American yard as to building it in a foreign yard. The subsidy rates then decline annually, it is hoped resulting in a similar decrease in costs in the American yard. The act also guaranteed most of the loan taken to cover shipbuilding costs and gave a tax incentive to ship operators if they set aside a portion of their revenues for future ship construction. This act and labor improvements have had an effect and by 1973 United States ship companies had increased their share of total United States construction to 6.4% from 4.6% in 1972. In 1974 the value of private shipbuilding in the United States reached $4.2 billion, up from $3.8 billion in 1973. The upward trend is expected to continue for these and other reasons. For example, only United States ships will be allowed to carry oil from the end of the Alaska pipeline in Valdez to United States ports. In addition, there are laws, called cabotage laws, that restrict foreign entry into United States ports or to foreign engagement in certain businesses within United States waters. (Actually, most countries have similar laws.) One law bans foreign fishing vessels from unloading their catch in United States ports; likewise foreign tugs may not tow United States vessels in United States waters.

Port and Offshore Facilities

Ports are extremely important facilities in the economic development and growth of a country. Although ships have continued to grow in size and complexity, there generally has not been a corresponding worldwide improvement in port and harbor facilities. Many major harbors, especially in the United States, are not deep enough to handle large vessels; Seattle, Washington and Long Beach, California are the only United States ports able to receive vessels up to 100 000dwt in size. Worldwide, in the early 1970s, there were only nine foreign ports which could receive ships larger than 200 000 dwt.

The principle method of handling large tankers is by offshore facilities. More than 160 such facilities are now operating in various parts of the world (Figs. 4-5 and 4-6). About 50% of these are in the Persian Gulf or Far East. At present, there are no single-point mooring systems off the United States, although several are planned.

Several LNG facilities are being built in various parts of the world. The development of a LNG plant is an expensive undertaking with at least $500 million needed for the terminal, connecting pipelines, and liquification plant. Because of the cost, the reserves of gas available to such a plant must be substantial to provide enough gas for a period of use of 20 years or more. Offshore complexes can be built to compress and move natural gas found in wells on the continental shelf. One such complex off Louisiana, called Stingray City, is situated in 160 ft of water and is about 100 miles from land (Fig. 4-7).

Offshore facilities generally have a single-buoy or multiple-buoy system,

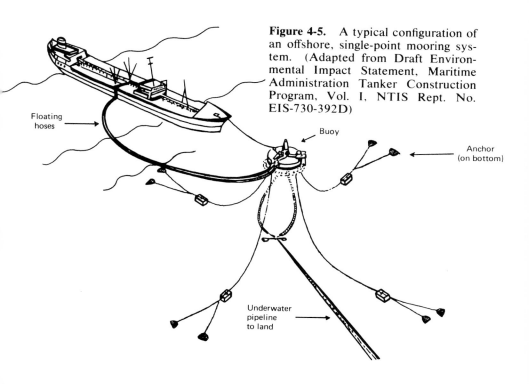

Figure 4-5. A typical configuration of an offshore, single-point mooring system. (Adapted from Draft Environmental Impact Statement, Maritime Administration Tanker Construction Program, Vol. I, NTIS Rept. No. EIS-730-392D)

Floating hoses

Buoy

Anchor (on bottom)

Underwater pipeline to land

Figure 4-6. A VLCC discharging its cargo at an offshore facility in the Caribbean. The oil is pumped ashore through a buried pipeline (see Fig. 4-5). (Photograph courtesy of the American Petroleum Institute)

anchoring space, and pipelines for transfer of oil or gas to or from near-shore facilities for refining or for storage (Fig. 4-5). The buoy is strongly anchored to the sea floor, and the tanker attaches or moors itself onto the buoy and discharges or receives its cargo. Most of the single-buoy or -point mooring

Figure 4-7. Stingray City, an offshore facility in the Gulf of Mexico that has the capacity to move 1 billion cubic feet of gas per day. (Photograph courtesy of Stingray Pipeline Company)

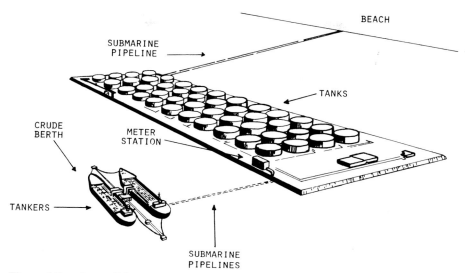

Figure 4-8. A possible configuration for an offshore island complex. (Adapted from Draft Environmental Impact Statement, Maritime Administration Tanker Construction Program, Vol. I, NTIS Rept. No. EIS-730-392D)

systems allow the attached ship to move with the wind, waves, or currents and to maintain the best possible position into the weather. The exact type of facility and its location will depend on general weather conditions, depth of water, distance from shore, construction costs, land facilities, and possibilities for environmental damage. Another type of facility is an artificial, offshore island complex (Fig. 4-8), with storage capabilities, but these can be extremely expensive.

Many of the arguments for offshore facilities are similar to those for the building of supertankers: Larger ships will be able to carry more oil more economically; larger ships mean less traffic, therefore less chance of collision, etc. The environmental arguments are not completely convincing. It is true that two 540 000-ton supertankers can carry as much oil as seventy 16 000-ton tankers (Fig. 4-9), and that possibilities of individual accidents may be reduced; but when they happen, more oil will be introduced into the environment (see Table 4-7). However, offshore facilities reduce traffic in harbors and shipping channels where most accidents occur. Likewise, it probably is easier to handle oil, with less risk of pollution, at an offshore port than at a land facility. In the design of any harbor facility—either onshore or offshore—or the evaluation of an existing one, it should be stressed that it is essentially impossible to have a pollution-free situation. Even if all design problems could be overcome, there would still be the possibility of human error.

One ancillary problem associated with the building of an offshore or

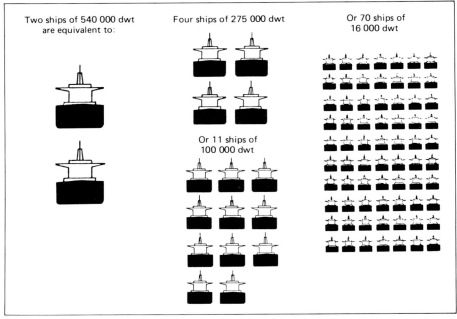

Figure 4-9.   A comparison of ships of similar carrying capacities. (From *Ecolibrium*, 1973, v. 2, n. 3)

onshore terminal is that it will attract numerous other industries. These can include refineries, petrochemical plants, and trucking or land shipping operations. This, in turn, can cause local concern both for the environment and for the pressure on the use of near-shore land.

For a country such as the United States there are four basic reasons that deep-water terminals will be built:

1. There is an economic advantage of using supertankers.

2. With so many supertankers already in operation, the present choice for the United States is either to build offshore terminals or to have the supertankers carry the crude oil to Canada, the Bahamas, or the Caribbean and then have it transported in smaller ships to the United States.

3. It is uneconomical to dredge existing harbors to the depths needed for supertankers, and disposing of the vast amount of dredged bottom material would present a major environmental problem.

4. The probable future increase in United States consumption of petroleum and the decreasing rate of domestic production means more petroleum will have to be imported.

A suggestion (Walsh 1976) along the lines of an offshore facility has been to "develop" a remote island in the Pacific where oil from large tankers

Table 4-7. Distribution of Number of Tankers, Port Calls, Accidents, Pollution-Causing Incidents, and Amount of Total Outflow in Percent by Four Major Oil Tanker Classes During the 1969–1974 Period[a]

Oil Tanker Class	Percent in Class	Percent of Port Calls	Percent of Accidents	Percent of Pollution-Causing Incidents	Percent of Total Outflow
2000 dwt to 50000 dwt	72.5	82.2	76.8	75.9	70.1
50000–100000 dwt	17.7	13.4	15.5	14.5	8.1
100000–160000 dwt	4.2	2.3	3.5	4.0	12.9
Greater than 160000 dwt	5.6	2.1	4.2	5.6	8.9

[a]From: An Analysis of Oil Tanker Casualties 1969–1974 (p.13).

could be taken and transferred to smaller tankers and then reshipped to such places as Japan. The island of Palau in the western Caroline Islands, which is within the United Nations Trust Territory of Micronesia and is administrated by the United States, is the prime candidate for this facility. Palau has a very fragile coral reef and lagoonal environment, but the program has strong economic appeal to this relatively poor island. The plan (sponsored by Japan and Iran) appears to have little chance for success at this time because improvements in the use of the Malacca Strait and elsewhere have decreased its appeal.

Problems Associated with Shipping

The most common problem associated with shipping is that of oil pollution. Tankers and other ships contribute to ocean pollution in several ways. These can include large-scale accidents or catastrophes such as the *Torrey Canyon* or *Amoco–Cadiz* spills; however, these are not the major ship pollution sources in the ocean. A more critical pollution damage comes from small leaks and spills, deliberate dumping of oil, and the discharge of bilge water as ships cross the ocean (see Table 6-6. and Fig. 6-4). Considering only the petroleum that enters the marine environment during shipping, the breakdown is about as follows: 50% from intentional cleaning and deballasting processes; 25% from pumping of bilges and during bunkering (fueling); 15% by accidents; and 10% while preparing for dry docking (Siler 1977). It is hard to determine the exact amount of oil introduced into the environment in these ways, but it is considerable and almost completely unnecessary. One estimate is that about 200 000 gallons of oil enter the ocean every hour of every day. Most of the United States accidents occur along its east coast, where traffic is heavy and the weather can be especially rough (Table 4-8).

Because much of the oil discharged into the oceans is deliberate, this pollution could be minimized by stronger regulations and better shipboard operations. Some accidents are clearly beyond control, although caused by human error, especially most ship groundings and collisions. Many sea-going oceanographers have had the experience of watching a supertanker or similar large ship pass by them at sea without being able to detect anybody on the bridge of the ship. Some of the conditions leading to such accidents are amazing. The book *Supership,* by Noël Mostert (1974), details some of these horrors; for example, in 1972 two Liberian registered supertankers, the *Oswego Guardian* and the *Texanita,* both about 100 000 tons, collided near Capetown. Each ship was traveling at full speed in dense fog and had observed the other on radar, but neither made any attempt to reduce speed or apparently to modify their course. After the collision one ship rapidly left the area without attempting to pick up survivors and their SOS message gave a wrong position. As a result of this avoidable accident, which killed 33

Table 4-8. Distribution of Accidents, Petroleum-Causing Incidents (PCI), and Outflow among Three Coasts of North America During the 1969–1974 Period, as a Whole and Only Those Within Coastal Zones[a]

Geographic Location	Total Accidents	Accidents Within 50 naut. mi. of Land but not Within Entranceway, Harbor, or at Pier	PCI's (total)	Pollution-Causing Incidents, 50 naut. mi. of Land but not Within Entranceway, Harbor, or at Pier	Total Outflow (long tons)	Outflow within 50 naut. mi of Land but not Within Entranceway, Harbor, or at Pier (long tons)
East coast of North America	430	62	70	13	195 042[b]	22 677[c]
Gulf of Mexico/Caribbean Sea	283	32	30	2	14 889	6 195
West coast of North America	78	14	13	1	9 228	1 350
Total	791	108	113	16	219 159	30 222

[a]From: An Analysis of Oil Tanker Casualties 1969–1974 (p. 16).
[b]Includes four catastrophic structural failures that occurred at sea, with total outflow of 110 360 long tons, and one explosion, which also occurred at sea, with total outflow of 31 716 long tons.
[c]Includes two catastrophic structural failures, with total outflow of 17 878 long tons.

men, one of the ship's masters had his license suspended for 18 months; the other captain (the ship that left the scene) had his revoked.

One of the worst areas for accidents is the English Channel. In an attempt to control accidents in this busy area, sea lanes or traffic lanes have been established for ship movement. An especially tragic series of incidents occurred in the Channel in early 1971. It started with a small Peruvian tanker entering one of the lanes but going in the wrong direction. It then hit and sank a Panamanian registered tanker, the *Texaco Caribbean*. British authorities quickly indicated the area of the wreck with two lighted buoys and an anchored lightship. The next day a German freighter hit the wreck and sank with considerable loss of life. This new wreck was marked with buoys, but within a few weeks a Greek ship hit one of the two wrecks with the loss of her entire crew. According to British authorities, 16 other ships in a 2-month period ignored the warning systems and entered, but survived, the wreck area.

In late 1975 a giant Norwegian freighter, the 227 556-ton *Berge Istra*, mysteriously disappeared in the Pacific. It was a record loss, because this ship was reportedly insured for $27 million. No indication of problems or distress was received from the ship, which was considered one of the safest in the Norwegian merchant fleet. Twenty days after the ship was reported lost, a Japanese fishing boat picked up two survivors from the vessel. They reported that they were painting the ship (which was carrying 180 000 tons of iron ore) when there were three explosions that tore the ship apart and sank it. That is all that is known.

It is estimated that every year about 6% of the world's ocean-going ships are involved in some sort of accident. This is about 1500 accidents per year. In the 1964–1977 period, 198 tankers were declared to be total losses because of fire or explosions. From 1969 to 1974 there were 3709 accidents involving tankers, of which 519 were pollution-causing incidents. According to a 1978 study (Analysis of Oil Tanker Casualties 1969–1974) the accidents were caused by: structural failures, 34%; groundings, 31%; collisions, 22%; and explosions, 13%. There is no doubt that some accidents could have been avoided with better seamanship and stricter regard to rules and international regulations and navigation.

Clearly a major problem with large ships is their safety. Many of these vessels have simply vanished, probably because of being cracked apart by large waves. The huge size of the ships means there are times when their bows and sterns may be supported by the crests of two waves, while the middle portion of the ship is relatively unsupported in the trough of the wave. This situation can result in considerable stress to the vessel. Another difficulty comes from the size of the load the ship is carrying—ships have load restrictions depending on the area of the ocean that they travel in. Ships in tropical seas can carry the heaviest loads, whereas ships in winter zones or colder seas are supposed to carry the lightest loads. Consider, however, a fully loaded tanker departing from the Persian Gulf (tropical area) and making a trip to the United States by going through the winter seas around the tip of Africa. To comply with safety regulations the tanker should unload oil when it reaches the cold waters or not be fully loaded when it left the Persian Gulf. The latter is rarely done; the former would involve pollution or a costly transfer; in addition either of these would involve loss of profit. In 1968 the maritime regulating body International Maritime Consultative Organization (IMCO, a United Nations agency) essentially eliminated these regulations. Ironically the waters near the tip of South Africa are also areas of strong seas, so the previously mentioned problem of ship support by waves becomes critical at the same time that the overloading of the ship does. It is not surprising that this area of the world is the site of numerous accidents to large tankers.

Operating ships under flags of convenience may contribute to accidents, because some countries permit operations without major safety inspections. This can result in many old, rusty, ill-kept ships operating when they would be best served by being junked. Another aspect of some flags of convenience is that the composition and quality of the crew can be kept at a minimum. For example, American tankers are required to carry at least nine licensed crewmen, whereas Liberian registered ships only require three. It can be seen from Table 4-9 that Greek and Liberian tankers have the worse safety records, whereas the United States is about average.

Liberian ships at the end of 1976 had an especially bad run of trouble. Within a 2-week period, the *Argo Merchant* went aground off Cape Cod on Nantucket Shoals and spilled 166 000 barrels (about 7 600 000 gallons) of

Table 4-9. Distribution of Accidents, Pollution-Causing Incidents, and Total Outflow as a Function of Fleet Size by Registry During the 1969–1974 Period[a]

Country	Percent Accidents/ Percent Total Ships[b]	Percent Pollution-Causing Incidents/ Percent Total Ships[b]	Percent Outflow/ Percent Total Ships[b]
Liberia	1.71	1.52	1.61
United Kingdom	1.23	1.18	0.64
Japan	0.17	0.38	0.51
Norway	0.57	0.74	0.52
Greece	2.34	2.07	1.41
France	0.52	0.45	0.06
United States	1.09	0.77	0.72
Italy	0.46	0.77	0.29
Panama	1.25	1.27	1.14
Sweden	0.68	1.11	0.68

[a]From: An Analysis of Oil Tanker Casualties 1969–1974 (p. 11).

[b]Values greater than one indicate higher (or worse than average) records relative to fleet size; less than one means lower or better than average records.

Table 4-10.   Global Traffic in Some Major Straits

Area	1969 Traffic (ships/day)	Estimated 1980 Traffic (ships/day)
English Channel	400	340[b]
Strait of Gibraltar	215	225
Cape of Good Hope	160	180
Coast of Japan	100	180
Malacca Strait	85	190
Persian Gulf	80	180

[a]Data from *Ocean Industry Magazine*, July 1973.
[b]Assuming that Suez Canal is not as important a commercial channel as in pre-1967 times.

No. 6 fuel oil. This was followed by the Liberian tanker, *Oswego Peace*, which spilled 2000 gallons into the Thames River, Connecticut. Finally, a third tanker, *Olympic Games*, ran aground in Delaware Bay and ruptured one of its fuel tanks, producing a 25-mile (40-km) long oil slick.

Part of the danger to the marine environment from shipping results from two important points: (1) More than half the seaborn trade is carrying petroleum products and (2) a large number of ships move daily in several narrow channels, thus increasing the possibility of collision. This latter problem will probably become even more critical in the future (Table 4-10). A major problem with supertankers and other large ships when traveling in narrow passages or channels is their mobility. It generally takes more than 3 miles (about 5 km) to stop one of these vessels when they are traveling at normal speed. This problem can be critical in shallow water, where they may be within feet of going aground (Fig. 4-10). As previously noted, an especially dangerous area for supertankers is the English Channel, where, besides the many ships, there are shifting sand ridges on the bottom which can ground an unsuspecting vessel. Over 3500 wrecks are known in this area, some of

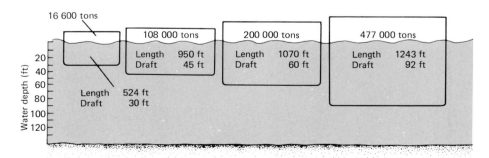

Figure 4-10.   General dimensions and drafts of some large tankers.

which have resulted from small ships trying to sail between the stern and bow lights of a supertanker, assuming that the lights belong to two separate ships. The presence of fixed platforms, stationary objects, or fishing gear in and near ship channels seems to have almost an inexplicable attraction for moving ships.

Because supertankers are so large, there are very few port facilities for these ships. If a problem develops, therefore, such as a leak caused by a minor grounding, there may be no readily available site for repairs. This has happened and ships have had to be towed thousands of miles until a locality for repairs could be found. Ships completely abandoned in international waters present interesting problems because they can be recovered and kept by their finders. This may lead to extremely profitable possibilities, especially if a large tanker splits in half; each half could contain oil valued in excess of a million dollars.

Tankers are not just dangerous because of the oil they carry. There have been instances when large tankers have been gutted by fire without any damage to the cargo. Especially dangerous is the vapor coming from the oil, which can explode or burn at very low temperatures. These vapors may also come from the thin film of oil that sticks to a cargo tank after it has been emptied. The cleaning of these tanks is therefore done with considerable care, usually with high-power water jets. Unfortunately, the water will then become a mixture of oil and water, which, if discharged into the sea, of course becomes a pollutant. The amount of oil that sticks to the walls of a large tanker can be considerable—as much as 1 or 2 tons of oil for a 200 000-ton tanker. Besides being dangerous for the ship, therefore, it presents a hazard to the marine environment. The cleaning operation is generally done while the ship is in transit back to its source of oil and is traveling under ballast (water is added to compensate for lack of cargo and to keep the ship properly and safely submerged). Several 200 000-ton ships have had explosions during cleaning, and at least one is known to have sunk.

The International Maritime Consultative Organization (IMCO), is essentially the principal international agency concerned with maritime jurisdiction, especially pollution. For any IMCO regulation to be accepted the member nations of IMCO have to agree to the regulation by internal or domestic legislation, which can be a very slow process. In addition, not all states belong to the organization and, finally, even if restrictive legislation concerning pollution or size of cargo is adopted, there are no effective methods of enforcement. In recent years IMCO conferences have adopted several new and important pollution prevention measures. For example, new crude carriers over 20 000 dwt must have protectively located, segregated ballast tanks; also radar and two remote steering systems will be mandatory on certain size ships. These regulations become effective when accepted by 15 states whose combined tonnage equals or exceeds 50% of the world's total.

Pollution by the water used for ballast and for the cleaning operation can be kept to a minimum by either of two methods. One is by direct off-loading of all water at a land base, where it can be treated and the oil can be removed and recovered. Very few of these facilities exist, but they would almost pay for themselves, once the oil and water are separated. The second method is by the so-called load-on-top technique (see Fig. 6-11).

The adoption of either one of these two techniques would go a long way toward solving oil pollution problems. According to various estimates, 1.5 million tons of oil are being discharged yearly into the sea from ballast operations and pumping of bilges. IMCO has made it an offense to discharge such oil within 50 miles (80 km) of land [100 miles (160 km) in some special instances]; however, punishment is only possible in the country where the ship is registered or if by chance the ship enters a port in the country whose waters it has polluted—and proof of the crime is again a different and very difficult matter.

Liquid natural gas (LNG) tankers are considerably safer at sea than oil tankers. This partially because of differences in construction; LNG tanks are generally circular and are not coincident with the outer hull of the ship as oil tanks are. If an LNG vessel is grounded it is probable that the cargo tank will not be damaged, whereas the opposite often occurs with an oil tanker. Perhaps most important is that LNG can quickly evaporate and essentially disappear into the atmosphere, whereas oil will remain much longer in the marine environment and can cause considerable environmental damage. There can be a major catastrophe, however, if LNG is released and ignites, especially if this occurs near a large population area. For this reason, movement of LNG tankers in coastal harbors is generally done under conditions of considerable care.

Ships can also cause an interesting type of biologic pollution resulting from organisms that may attach themselves to a ship's hull and are transported and settle in an area where such organisms had previously not been found. This can also occur if species, especially planktonic forms, get in the ballast water of a tanker and survive when this water is discharged into a new environment. In this manner organisms can traverse otherwise effective ecologic barriers, such as the Suez or Panama Canals. There have been some interesting examples of this type of biologic pollution, such as the finding of blue crabs common to the waters of Chesapeake Bay in the near-shore waters of Egypt and Israel, or the presence of European and Atlantic species of eels in San Francisco Bay. The introduction of such forms can cause ecologic havoc, because the normal feeding and living patterns of the indigenous fauna can be severely disrupted by a new and rather dramatically introduced species.

In summary, the problems of oil pollution by ships at sea could be reduced by three main actions:

1. Clear and positive control of ships, especially near or within coastal waters

2. Regulation to control accidental spills, including establishment of responsibility and liability, and strict regulations and enforcement against the almost routine dumping of bilge and ballast water without first separating out the oil it contains

3. Improved ship design

There have been some positive actions taken along these lines and the future seems favorable. It is also probable that the Law of the Sea Conference (see Chapt. 3) will produce regulations that lead to safer operation of ships. The next section discusses some potential problems for shipping that may arise from the conference.

Shipping and the Law of the Sea

The shipping industry could be affected by restrictions arising from the Law of the Sea Conference or by unilateral extensions to territorial seas. This can occur in two ways: The first is by the extension of the width of territorial seas from the present 3 naut. mi. to 12 naut. mi., which could impede free passage through what are now international waterways (i.e., those narrower than 24 naut. mi. but wider than 6 naut. mi.).

A strait can be defined as a narrow body or passage of water between two land masses that is less than or equal to the width of the territorial sea claims of the countries bordering it and that connects with the high seas or unrestricted waters. For example, if two states opposite to each other across such a passage each had accepted a 3-naut. mi. wide territorial sea, the width of the water would have to be less than 6 miles to be legally considered as a strait. If it were greater than 6 miles in width, it would, under these conditions, be considered an international waterway.

The changing of international waterways to straits because of expansion of the territorial seas would have three immediate effects:

1. The movement of warships through these straits would be restricted.
2. The previous right of free passage would be replaced by rights of innocent passage.
3. Rights of unrestricted overflights by airplanes would be lost.

If territorial seas were extended to 12 naut. mi., over 100 straits would be changed from international waterways to coastal state control. Of these, 32 are relatively important and they can be divided into four main groups. The largest group includes 18 straits that would fall under the control of maritime states, who would probably not endorse any severe restrictions to shipping (Table 4-11). Of the remaining, six are relatively minor and insignificant, five are important but have longer alternate routes, and three are important with no alternatives. Any restrictions on the latter can affect international

Table 4-11.   International Straits That Would Be Affected by an Extension of Territorial Seas to 12 naut. mi.[a]

Those Under Control of Maritime States
Bonifacio, Chosen, Cook, Domenica, Dover, Foveaux, Gibraltar, Juan de Fuca, Kaiwi, Karpathos, Kithera, Martinique, Messina, Minorca, Oresund, St. George, St. Lucia, St. Vincent

Those Not Under Control of Maritime States

Minor Importance	Important but with Alternates	Important but with No Alternates
Dragon's Mouth	Malacca	Bab al Mandab
Hai-nan	San Bernadino	Dardanelles
Magellan	Singapore	Hormuz
Palk	Sunda	
Serpent's Mouth	Surigao	
Zanzibar		

[a]Adapted from Prescott (1975).

commerce. The three straits without alternatives, Bab al Mandab, the Dardanelles, and Hormuz, control access, respectively, to the southern Red Sea (and Suez Canal), the Black Sea, and the Persian Gulf—all very important shipping and military areas. There are some military restrictions on the Dardanelles. The Strait of Hormuz is presently under the control of the Iranian Government and political turmoil in that country may cause some problems for transportation of oil out of the Persian Gulf.

Merchant ships generally have the right of innocent passage through territorial waters but conceivably this could be withdrawn because of political or pollution reasons. The definition of "innocent passage" is not clear; neither is there international agreement on this point.

Some countries have imposed restrictions on straits under their control. One of the most important of these was the 1972 decision of the Indonesian and Malaysian Governments, whose territorial waters cover the Malacca Strait. They decided that tankers of greater than 200 000 dwt toms could not pass through the Strait fully loaded[2] and completely forbade the passage of ships greater than 500 000 dwt tons. Later in 1973 the South African government and IMCO required that loaded tankers keep at least 12 miles off the Cape of Good Hope.

[2]In early 1975, a 237 000-ton Japanese tanker, *Showa Maru*, went aground in the Malacca Straits, causing an exceptionally large oil spill.

There are four major canals that presently can be considered as international waterways. These are the Panama, Suez, Kiel, and Corinth Canals. Actually, only the Panama and Suez Canals are important for international trade and, if restricted, would require additional travel distance. The recent increase in tanker size, however, has diminished the economic potential and value for the Suez and Panama Canals, which are not wide enough to permit passage of these ships.

The second problem that could arise from the Law of the Sea Convention and the establishment of a 200-mile economic resource zone is that coastal states could impose unreasonable navigation or pollution requirements on ships that pass through their zones. These requirements could change from country to country, although it would be hoped that international conferences could lead to international standards. As previously mentioned in the case of the Malacca Straits, states can assert that certain classes of ships, such as oil tankers or nuclear-powered ships, are inherently noninnocent. The full effects of a Law of the Sea Convention on shipping could result in major changes in the shipping industry.

References

Alexander, L.M., and T.A. Clingan Jr. (Eds.), 1973, *Hazards of Maritime Transit.* Cambridge, Mass.: Ballinger, 1968 p.

Aron, W.I., and S.H. Smith, 1971, Ship canals and aquatic ecosystems: *Science,* vol. 174, p. 13–20.

Bragaw, L.K., H.S. Marcus G.C. Raffaele, and J.R. Townley, 1975, *The Challenge of Deepwater Terminals.* Lexington, Mass.: D.C. Heath and Co., 162 p.

Deepwater ports: Issue mixes supertankers, land policy: 1973, *Science,* v. 181, p. 835–837.

*Draft Environmental Impact Statement, Maritime Administration Tanker Construction Program, Vol. 1, NTIS Rep. No. EIS-730-392D.

Hodgson, R.D., and McIntyre, T.V., 1973, Maritime commerce in selected areas of high concentration: *In* Alexander, E.M., and Clingen, T.A. Jr., Eds., *Hazards of Maritime Transit.* Ballinger Publishing Co., Cambridge, Mass, p. 1–18.

Icebergs and oil tankers: USGS glaciologists are concerned: 1975, *Science,* v. 190, p. 641–643.

Jantscher, G.R., 1975, *Bread Upon the Waters: Federal aids to the Maritime Industries,* Washington, D.C.: The Brookings Institute, 164 p.

*Lloyds Register of Shipping Statistical Tables, 1974.

Knight, G.H., 1973, International legal problems of offshore port facilities: Marine Technol. Soc. Jour., v. 7, p. 4–12.

Marcus, H.S., 1974, Maritime transportation systems: *In* Kildow, J., Ed., *International Transfer of Marine Technology: A Three-Volume tstudy.* Cambridge, Mass.: Massachusetts Institute of Technology, Rep. No. MITSG 77–20, September 1977, p. 81–142.

*References indicated with an astrisk are cited in the text, others are of a general nature.

Maritime Administration Reports (1976, 1977). United States Department of Commerce.

*Mostert, N., 1974, *Supership*. New York: Alfred A. Knopf, 332p.

*Organization for Economic Cooperation and Development, 1978, *Annual Report for 1977*. Paris, France.

*Prescott, J.R.V., 1975, *The Political Geography of the Oceans*. New York: John Wiley and Sons, Halsted Press, 247 pp.

*Siler, O.W., 1977, The suppression of oil spills: *Ecolibrium*, v. 6, n. 3, p. 2–3.

United Nations Conference on Trade and Development, *Review of Martitime Transport*, 1975, Geneva, 101 pp.

United Nations Conference on Trade and Development, The Secretariat, 1972, *Shipping in the Seventies*. Geneva, 42 pp.

United Nations Secretary-General, 1972, *United Nations Economic and Social Council Uses of the Sea*, New York: E/5120, 28 April 1972.

United States National Oceanographic and Atmospheric Adminstration/Environmental Protection Agency, 1978, *AMOCO-CADIZ Oil Spill: A Preliminary Scientific Report*. NOAA/EPA Spec. Rept. Washington, D.C., 347 pp.

*United States Office of Technology Assessment, 1978, *Analysis of Oil Tanker Casualties 1969–1974: 1978*, A National Ocean Policy Study (prepared for the use of the Committee on Commerce, Science and Transportation), United States Government Printing Office.

*Walsh, J., 1976, Superport for Palau debated: Ecopolitics in the Far Pacific: *Science*, v. 194, p. 919–921.

Chapter 5

The Resources of the Ocean[1]

Introduction

Resources can be of two basic types: those, such as food, that are renewable because they can be replaced by the growth of plants and animals, and those, such as most mineral resources, including coal and oil, that are depletable and so are nonrenewable because there is a fixed supply. Both renewable and nonrenewable resources are present in large quantities in the ocean and there is considerable enthusiasm among the countries of the world to obtain them. With regard to the biologic or renewable resources, some feel that the present fish catch from the sea may already be dangerously close to exceeding the maximum sustainable yield of this resource. Little, however, is really known about the distribution of fish and other organisms in the ocean and how they can be best managed. A new approach, aquaculture, will allow some fishery products to be harvested similarly to crops on land, thus changing the fisherman from a hunter to a farmer with a substantial increase in efficiency. Other potential resources, such as energy from the ocean, are discussed in Chapter 9.

Within recent years considerable emphasis has been given to the mineral resources of the ocean, especially oil and gas from the sediments of the continental margin. At present about 20% of the world's petroleum products come from the sea and this may reach 50% by the year 2000. Manganese nodules, which cover a large part of the deep-sea floor, are also thought to be a valuable ocean resource.

[1]Parts of this chapter are adapted from Ross (1977).

Table 5-1. Some Possible Marine Mineral Resources and Their Source.

Mineral Resource	Source
Boron, Bromine, Calcium, Magnesium, Potassium, Sodium, Sulfur and Uranium	Sea Water
Sand and Gravel, Phosphorite, Glauconite, Lime and Silica, Sand, Heavy Minerals (Magnetite, Rutite, Zircon, Cassiterite, Chromate, Monazite, Gold)	Sediment (Continental Shelf and Slope)
Copper, Lead, Silver, Zinc	heavy metal muds
Oil, Gas, Sulfur	Subsurface (Continental Shelf, Slope and Rise)
Manganese nodules (Copper, Nickel, Cobalt and Manganese)	Deep Sea

The United States is especially concerned with minerals from the sea, because of its dependency on foreign sources for many strategic minerals. Seventy-one different materials and minerals are generally considered to be critical for a modern industrial state; of these, the United States imports 69 either in part or completely. The Soviet Union, in contrast, is presently self-sufficient in all but two of these resources. Many of these 69 minerals are found in considerable supply in the ocean (Table 5-1).

It is, however, somewhat ridiculous to consider that the world is running out of minerals, because the entire planet is composed of minerals. The main exception to this is oil and gas. The amount of minerals in the earth's crust is awesome. For example, a single cubic kilometer of crustal rock contains as much as 200 000 000 tons of alumina and 100 000 000 tons of iron; a similar calculation can be made for sea water. This is not to suggest that these things be economically mined but that a physical exhaustion of mineral supplies with the exception of oil and gas is technically impossible if we are willing to pay the price of obtaining these resources.

The mining of minerals or drilling for oil within an area also used for the catching of fish or shellfish or for other purposes can cause environmental problems (see Chapt. 6). This potential conflict of interest presents a unique challenge for our decision makers that will be hard to evaluate correctly because of the scarcity of scientific information concerning marine animals, and the effects of pollution and conflicting evidence concerning how safe mining operations are. The wise use of the resources of the sea will be one of the major challenges for humankind over the next few decades.

The following sections of this chapter describe first the nonrenewable mineral resources of the ocean. This is then followed by a discussion of the biologic resources, including a brief section on aquaculture. Innovative energy sources from the ocean are considered in Chapter 9.

Mineral Resources of the Ocean

During the last few years a considerable interest has developed among marine scientists and individual countries in the mineral wealth of the ocean. Some recent estimates of the value of these marine resources have been staggering—values in the billions or even trillions of dollars have been stated. For some resources these estimates are ridiculous; for others they may turn out to be conservative. The most valuable marine mineral resource is hydrocarbons, such as oil and gas, which can be found buried within the sediments of the continental shelf and possibly also in parts of the continental rise.

The main problem in assigning a value to individual marine resources comes simply from the fact that they are under water and conventional techniques of mineral exploration and evaluation therefore cannot be used. Marine surveys, sampling, extraction, and refining are more expensive and complicated than those on land. Operations in the marine environment, besides providing occasionally inhospitable conditions to humans, will also take their toll on equipment and ships. Pollution control and unresolved legal questions concerning the ownership of the sea floor are complications that can add to the cost. Many potential marine resources must await technologic advances before they can be exploited, whereas others may never be developed because of their great distance from an area of consumption, which makes the transportation costs excessive, or because cheaper competitive land resource is available. Even with these problems it is not surprising that people are looking to the sea as a source of mineral resources. The present industrial growth rate of the world cannot continue to be met just by the land, which constitutes only 28% of the surface area of the planet. Occasionally the mineral shortages of the world are exaggerated and some minerals, potentially also available from the ocean, are actually very plentiful on land, although sometimes concentrated in only a few countries.

Besides finding new mineral sources there are other ways that we can increase our mineral base; these include improved mining, concentration and processing techniques; recycling of minerals through better conservation techniques; and substitution of more abundant minerals or artificial components, such as plastics, for those minerals in short supply.

The mineral resources of the ocean can be divided into four main categories: (1) those elements dissolved in sea water; (2) those minerals recoverable from the underlying bedrock, such as coal or iron deposits; (3) those minerals found on the ocean bottom; and (4) those minerals, such as oil and gas, within marine sediments (Fig. 5-1). The process of formation, extraction techniques, and economic potential vary considerably both between and within the categories. For example, the minerals found on the ocean bottom (the third category of the above) can be of five general types:

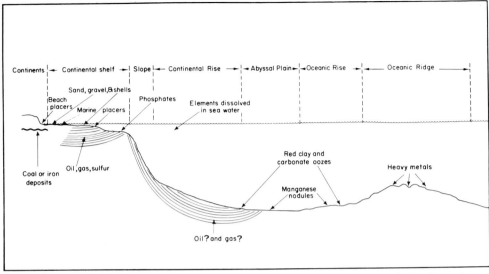

Figure 5-1. A generalized view of the main mineral resources available from the ocean floor. (Ross 1977)

1. Minerals or sediments that normally constitute the sea floor, such as sand and gravel in shallow water and carbonate oozes and red clays in the deep sea

2. Minerals that are concentrated by the actions of waves or currents, such as placer deposits

3. Minerals formed by chemical precipitation from the sea water, such as manganese nodules, or from volcanic processes

4. Deposits formed by biologic activity, such as reef rock, shells, or corals

5. Deposits of preexisting minerals that are now exposed by the eroding forces of the sea.

I have discussed the minerals of the ocean mainly by considering the area from which they come: continental margin (including coastal zone) or deep sea. Because many resources are common to both the coastal zone and the continental margin, they are included together. Among the reasons for this division are that a major consideration for any deposit is its distance from the land and therefore its ownership. Based on the probable results of the Law of the Sea negotiations, resources falling within a zone of 200 naut. mi. off a country's coast will be the property of that country (see Chapt. 3 for further discussion). There is also the possibility that the resources of the deep sea, in particular manganese nodules, will be mined at least partially under some form of international authority.

Mineral Resources of the Continental Margin

Included within the continental margin is the coastal zone, the continental shelf, the continental slope, and, if present, the continental rise or continental borderland. The continental margin contains an area equal to about 50% of the total land area of the world (Table 2-4). It is the transition zone between the continent, with its thick crust composed mainly of granitic rock, and the ocean, with its thin basaltic crust. The shelf and rise are usually areas of extremely large sediment accumulations, sometimes several miles thick.

The mineral resources of the continental margin can be divided into the following categories: elements in solution, including sea water; minerals recoverable from the underlying bedrock; minerals on the ocean bottom; and oil, gas, and sulfur.

Sea Water and Elements in Solution

Sea water is the most abundant fluid on this planet, but fresh water is in short supply in many localities. Fresh water is a marine resource and many salt water conversion plants have been built or are planned in areas where there is a shortage. The need for water may increase even more because of energy shortages. Considerable water would be needed in processes that produce synthetic natural gas from coal or in oil shale development. A plant producing one million barrels of oil (42 million gallons) per day from oil shale would use about 100–170 million gallons of water per day. These estimates clearly indicate that severe shortages of water are possible in the near future.

At present there are over 700 desalinization plants in the world that have capacities of 25 000 gallons or more per day of fresh water. Even more will be built as technology is improved so that the cost is competitive with other sources of fresh water (see Chapt. 9 for a discussion of various desalinization techniques). In such areas as the Persian Gulf, fresh water is in such short supply that water produced by the desalinization of sea water is the main source. Some imaginative ideas have been developed for sources of fresh water, such as towing icebergs either from the Arctic or Antarctica to an area where they could be used (pages 298–299). Another innovative idea is to extract water vapor from the atmosphere in tropical areas by cooling it with cold ocean water brought from depths to the surface (see pages 287–288).

Sea water probably contains all elements known on land, although only about 80 or so have been detected. Common salt, bromine, and magnesium are the main elements presently being commerically removed from sea water. There is considerable potential, however, for other elements if the technology and economics should improve.

The water in the ocean contains a virtually inexhaustible supply of many elements. Considering its volume—about 1350×10^6 km^3 or 318 million cubic miles—and its salinity—about 35°/₀ or 3.5%—each cubic mile of sea water contains about 165 million tons of dissolved elements. The ocean contains over 5 billion tons of uranium and copper, 500 million tons of silver, and as much as 10 million tons of gold. For gold this is about 5 lb for each person on earth, or about $24 000 at $300 per ounce. These estimates are not very realistic when the cost of extraction of the elements are considered. For example, the average concentration of gold ranges from 0.000004 to 0.000006 ml/l or about 50 lb per cubic mile of water. This works out to be about a thousandth of a penny per ton of sea water. Other elements, such as bromine and magnesium, have more reasonable values and can be profitable recovered from sea water.

Minerals Recoverable from the Underlying Bedrock on the Continental Margin

This includes land-based mining operations, such as coal mining in Japan, the United Kingdom, and Nova Scotia, that have been extended out under the sea. In most instance, these operations are similar to that of land mining and would not have started if the land portion of the deposits had not already been known. An example is a bauxite deposite discovered in 1955 on the shore of the Gulf of Carpentaria, Australia that has been extended seaward and presently is one of the world's largest bauxite mines.

Phosphorite-rich rocks and sands are exposed on some areas of the sea floor and could be mined and used for fertilizer (Fig. 5-2). In spite of the fact that these deposits usually assay between 20% and 30% P_2O_5, a value close to that needed in fertilizer manufacture, their water depth and problems of recovery, transportation costs, and more competitive land sources generally prevent most marine phosphorite from being mined at present. Marine mining operations are being seriously considered, however, off some offshore localities near agricultural areas, especially if the countries are phosphorite importers. It has been estimated that near-future needs for P_2O_5 will increase by about 6%–10% per year. From 1965 to 1975 there was a 100% increase in demand that reached a total of over 120 million metric tons of phosphate rocks a year. Phosphate deposits off southeastern United States cover an area of perhaps over 125 000 km² and contain several billions of tons of phosphate, although the grade of the P_2O_5 in these deposits is lower than that of most land reserves. Pollution and other problems involving the land mining of large amounts of phosphoritic rock may also make marine sources of phosphate more attractive in the future.

Phosphorite deposition generally occurs in areas of upwelling, where cold nutrient-rich waters (phosphate is a nutrient) rise to the surface. As these waters are warmed and their pH increases, phosphate is precipitated. If the precipitates are not diluted by land-derived material, a rich phosphate deposit may occur; topographic highs near cold currents are especially favorable areas. Probably many marine phosphate-rich deposits are still to be discovered.

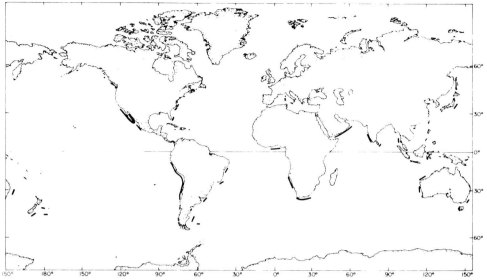

Figure 5-2. Offshore areas (black "bars") where phosphorite has been found. (McKelvey and Wong 1970)

Mineral Deposits Found on the Ocean Bottom in the Continental Margin Region

Most ocean-bottom mineral resources on the continental margin are related to three main factors; first, the large amount of sediments and second, the reworking effect of waves and currents combined with changes in sea level caused by recent ice ages. A third factor, discussed later, is how the continental margin has evolved throughout geologic time, including the processes of sea-floor spreading and plate tectonics. Many of the world's continental margins have a considerable thickness of sediment on their shelves, generally less on the slope, and sometimes a thick mass of sediment near the base of the slope forming the continental rise. These sediments derive from the large amount of material eroded from the continent and carried to the ocean by rivers, wind, and coastal erosion. In some regions sediments have been accumulated for several 100 million years and may be more than 10 km thick. These sediments generally have a relatively high content of organic material because of life processes in the overlying water during their deposition and this, combined with their thickness, is favorable for oil and gas accumulation. This aspect is discussed in more detail in a following section.

The rise and fall of sea level over the past million years is the second factor that has influenced mineral resources on the continental margin. These changes in sea level were in response to extensive periods of glacial growth and melting, the most recent glaciation occurred about 15 000 years ago

(Fig. 2-3). During this glaciation enough water was removed from the ocean and incorporated into the glaciers to result in a drop in sea level of 130 m (over 400 ft). As the glaciers melted, during the past 15 000 years, the water returned to the ocean and sea level slowly rose. Therefore, 15 000 years ago sea level was 130 m lower than at present. With the melting of the glaciers the shoreline migrated landward over the previously exposed portion of the continental shelf; the shoreline would have moved seaward when the glaciers were forming. As the shoreline moved so did the beaches and other near-shore features. If the movement of the shoreline was very fast, a beach could be stranded, forming an ancient beach or ridge. Associated with a shoreline are breaking waves and near-shore currents (Fig. 8-6) that can remove fine-sized material from the sediment and carry them seaward, leaving a coarser-grained beach or sand deposit. With a narrow shelf having a modest slope, the migration of the shoreline would be relatively slow and the reworking (by waves and currents) could be more intense, resulting in well-sorted (sediment particles generally of the same size) beach sands over the continental shelf (Fig. 5-3a). With a wide, gently sloping shelf, the shoreline would move more rapidly per unit time, and reworking would not be as intense (Fig. 5-3b). If the rise or fall of sea level stopped at a particular level for a period of time, it is also possible that most of the lighter sand particles could have been carried seaward, leaving a deposit enriched in the relatively heavy grains. This type of deposit, called a placer, may be very valuable depending on the minerals it contains.

Placers can contain economic concentrations of heavy minerals, such as rutile, magnetite, ilmenite, zircon, platinum, and even gold. Although beaches are where the reworking can be most intense, and placers are common, most modern beaches are limited in extent and generally are better used as recreational areas, so they are rarely mined. Offshore beaches or

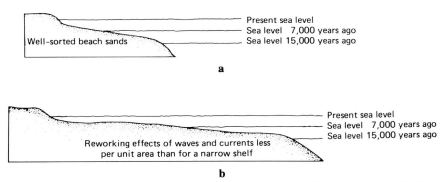

Figure 5-3, a and b.   The relative effects of rising sea level on: **a** a narrow shelf and **b** a wide shelf.

ridges therefore are generally more acceptable to mining; they also tend to be fairly numerous and extensive. They can be found by using conventional surveying and sampling techniques. Exploration for offshore placers generally starts with a study of the geology of the adjacent land, because it is these rocks that are the source of the minerals in the placers. If such minerals are found on land, the probability of offshore deposits are improved. Offshore exploration usually involves seismic and bathymetric surveys to delineate old or buried beaches and buried river channels. Estimates of sediment thickness combined with mineralogic analyses of sediments collected by coring or drilling are then used to evaluate the deposit. Most offshore placers will contain the mineral magnetite, which is magnetic, and high concentrations of it can be detected with a magnetometer. The more common minerals found in beach or offshore placers are listed in Table 5-2.

Placer deposits may also be found in river valleys, both present ones and ancient ones on the shelf, where again the reworking effects of water, in this case from the river, has concentrated the heavier minerals and removed the lighter. Examples are gold and platinum placers off Alaska, tin and rutile off Australia, and diamonds off Southwest Africa. If a river valley has mineable deposits near the ocean it is probable that the offshore parts of the river also have similar deposits.

The second most valuable marine mineral deposit after oil and gas is sand and gravel. Although not as exotic as other marine commodities, its extensive occurrence, ease of mining, and need in the building industry have made it a most valuable source. On a weight basis, it is one of the most used minerals. In the United States over 900 million tons of sand and gravel valued at $1.6 billion were mined during 1974 and most estimates are that need will at least double by the year 2000. It generally is not realized how extensively sand and gravel are used; for example, a 30 × 40 ft (9 × 12 m) concrete basement for a house would use 80 tons of sand and gravel aggregate, and 1 mile (1.6 km) of a four-lane highway would use 60 000–100 000 tons of mineral aggregate.

Although sand and gravel are low-cost commodities, their value will increase in the future as present sources on land are used up or covered by building in coastal areas. Because transportation is a major cost in a sand or gravel operation, it is obvious that there will be increased pressure to mine the vast deposits of sand and gravel in the marine environment that in many instances are very close to the area of use on land. Manheim (1972b) has evaluated sand reserves off the northeastern United States and concluded that the upper 3 m of sediment contain more than 400 billion tons, or several 100 years' of supply. The quantity of material dredged from the United States continental shelf has doubled during the 1960–1970 period to some 50 million cubic yards of sand and gravel and 20 million tons of oyster shells per year. Even larger amounts are mined off other countries; especially active is Great Britain. At present, 32 British companies are operating over 75

Table 5-2. Mineral Deposits on the Sea Floor of the Continental Margin

Minerals	Use	Possible Mineable Marine Areas[a]	Value[b]
Marine placers			
Gold	Jewelry, electronics	Alaska, Oregon, California, Phillippines, Australia	$350 or more per ounce
Platinum	Jewelry, industry	Alaska	$500 or more per ounce
Magnetite	Iron ore	Black Sea, Russia, Japan, Phillippines	$6–11/ton
Ilmenite	Source of titanite	Baltic, Russia, Australia	$20–35/ton
Zircon	Source of zirconium	Black Sea, Baltic, Australia	$45/ton
Rutile	Source of titanite	Australia, Russia	$100/ton
Titanite	Source of titanite	Australia, Phillippines	
Cassiterite	Source of tin	Malaysia, Thailand, Indonesia, Australia, England, Russia	
Monazite	Source of rare earth elements	Australia, United States	$170/ton
Chromite	Source of chromium	Australia	$25/ton
Sand and gravel	Construction	Most continental shelves	
Calcium carbonate (aragonite)	Construction cement, agriculture	Bahamas, Iceland, southeastern United States	
Barium sulfate	Drilling mud, glass, and paint	Alaska	
Diamonds	Jewelry, industry	Southwest Africa	
Phosphorite	Fertilizer	United States, Japan, Australia, Spain South America, South Africa, India, Mexico	$6–12/ton
Glauconite	Source of potassium for fertilizer		
Potash	Source of potassium	England, Alaska	

[a]Not necessarily including all areas, because in many localities exploration has been nil or minimal.
[b]Can vary depending upon degree of refinement; data from various sources and may be in error because of changing economic conditions.

dredges and mining sand at a cost of between 35 and 50 cents per ton, a cost considerably less than on land. Marine deposits are especially valuable if they are near the locality where they will be used, because transporting the material by truck over a distance of about 25 miles or 40 km essentially doubles the cost of the sand and gravel.

Dredging of sand, gravel, or shells is fairly easy with existing technology (Fig. 5-4), providing that the water depth is reasonable and storms and waves are not excessive. It can be possible to combine operations for mining both sand and gravel and heavy mineral placers using the same dredging operations. There can be environmental problems associated with offshore dredging; however, in some instances these may be less than those from land quarries near residential sections, where large trucks are needed to transport the material to the area of usage. Offshore problems involve erosion of

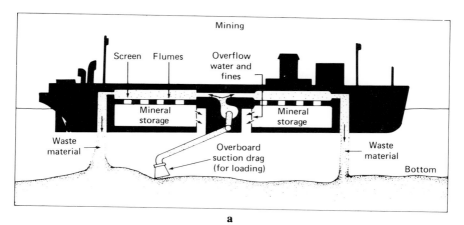

a

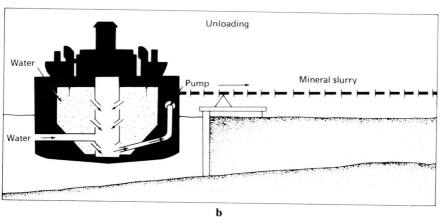

b

Figure 5-4, a and b.   Cross sections of a dredge used for mining sand and gravel. (Courtesy of Construction Aggregates Corporation, Chicago, Illinois)

beaches, because offshore sand may be used in the seasonal beach cycle, and interference with bottom ecology. The actual dredging of material can also introduce large quantities of suspended material into the water, which can have environmental effects on floating organisms and later on bottom-living organisms when the material settles out.

Other marine mineral resources on the continental margin include calcium carbonate, barite, and, in the past, diamonds off South Africa. Although the diamond mining received considerable publicity, it apparently never was profitable because operations were hampered by extreme sea conditions and relatively low concentrations of diamonds. Calcium carbonate is sometimes mined and used as a source of lime for agriculture or for cement. In most situations, it is easier and cheaper to quarry onshore sources.

Barite, used by the petroleum industry as a major ingredient in drilling sand, is found as concretions or small crystals of barium sulfate on some areas of the sea floor. It is presently mined off the Alaskan coast at the rate of about 1000 tons per day, but its main source is land.

One other potential mineral that may be mined is glauconite, an iron- and potassium-rich clay mineral. Glauconite is probably directly precipitated from sea water and generally accumulates on the continental slope or other shallow areas that have low sedimentation rates. Glauconite contains between 4% and 9% K_2O and may be used as a source of potassium for fertilizers.

Before I discuss oil, gas, and sulfur, which are the principal resources of the continental margin, it is appropriate to consider briefly the concept of sea-floor spreading, which explains why some resources are in one locality and not in another.

Sea-Floor Spreading and Mineral Resources

One of the most exciting scientific ideas of recent years is the concept of sea-floor spreading. The basic aspect of this concept is that new sea floor forms along the midocean ridges and slowly moves away from the ridge, eventually to either be part of an expanding ocean floor or to be consumed or subducted at the edge of the continents in a trench (Fig. 5-5). More recently, it has been shown that the movement can be visualized as several large plates (Fig. 2-9) moving across the earth's surface at rates of a few centimeters per year. This aspect, called plate tectonics, and other parts of this concept have been discussed in more detail in Chapter 2 (pages 14–17) and are not treated in detail here except where they affect mineral resources.

One interesting aspect of the concept is that mineralization can occur along some of the midocean ridges as the new sea floor is created. This mineralization can form deposits of copper, zinc, iron, cobalt, manganese, and other minerals. One of the most dramatic examples of this type of deposit is the heavy metal sediments of the Red Sea, described in a later sec-

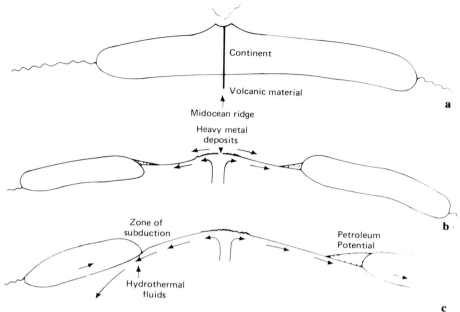

Figure 5-5.   The breaking apart of a continent because of sea-floor spreading. In the first phase (a), an initial large continental mass is intruded along zones of weakness by volcanic material. The continental mass eventually breaks apart (b), forming two continents. During this phase and the next, heavy metal deposits can be formed along the ridge axis by volcanic activity. Note here that both continents are coupled to the sea-floor movement. A new ocean forms between the two continents and sediment is deposited along the margins of the continents. Sometimes the small ocean basin can become evaporitic, resulting in thick sequences of salt and anhydrite deposits. In the last phase, (c), the sea floor is underthrust, the continent on the left-hand side of the figure forming a zone of subduction and removing the sedimentary material either by carrying it under the continent or by accreting it onto the continent. This is similar to what is happening on the west coast of South and Central America and much of the western Pacific. On the right-hand side of the figure the continent is still coupled to the movement of the sea floor, and sediment accumulation on the continental margin continues. This side is similar to the history of the margins around the Atlantic Ocean and the east coast of Africa. (Ross 1977)

tion of this chapter. Similar types of deposits, although not as enriched, have been found on the ridges of the Indian, Pacific, and Atlantic Oceans. Our knowledge concerning the processes forming these minerals is incomplete and it is not known whether the mineralized areas ever can be mined. Even though the initial mineralization occurs at the ridge axis, some deposits may be found throughout the sea floor because, with time, this material at the ridge axis moves slowly across the ocean floor; it may eventually become accreted onto a continental land mass or be depressed or subducted below it (Fig. 5-5). Within the subduction zone there can be mineral enrichment, by rising hydrothermal fluids (hot fluids), and deposits rich in the previously mentioned metals may be accreted onto the continental plate to form part of the land mass.

There are basically two mechanisms of mineral deposit formation on oceanic ridges. The first occurs when the molted basalt starts to cool: Because of gravity (the crystals are heavier than the molten basalt), as the initial crystals form they settle down into the fluid from which they have crystallized. Subsequently forming crystals also sink, resulting in a sequence of layers of different mineral types. In this manner, elements that have been present in the original magma in only small quantities may be concentrated into extremely rich layers that eventually can be mined. Some of the ancient copper and nickel deposits in Canada and South Africa are thought to have formed in this manner. The other mechanism for forming minerals on oceanic ridges is metamorphism, or chemical alteration of the basalt as it cools in sea water. This process has formed the copper deposits found on Cyprus, an area that was probably once a spreading center.

As the ocean crust is thrust underneath the continents, mineral deposits may be formed by a complex process involving the melting and fractionation of the rocks as they are being dragged down. The process, although poorly understood, has led to increased prospecting activity on land in regions where ancient subduction or down-thrusting zones are thought to have occurred. It has been noted that many sulfide deposits are found along present or past plate boundaries where parts of the ocean basin have been thrust under a continent.

The sea-floor spreading concept can also be used to extrapolate backward and predict what the position of the continents has been at different times in the past (Fig. 2-10). If a mineral deposit that predates the actual separation of two then joined continents is now found near the edge of one of the continents, it is possible that part of the same deposit also may be found on the other continent. In other words, the original deposit may have been split in two. Such understanding could be of considerable advantage in exploring for new mineral deposits. For example, the locations of manganese, iron, gold, tin, uranium, diamond, and other mineral deposits are better known in Africa than in South America. In the past, however, these continents were joined, and recent prospecting in northeastern Brazil, the Guianas, and southern Venezuela and other areas assumed to be extensions of the African mineral belts has disclosed some similar deposits.

In some areas sea-floor spreading has occurred without subduction; here the crust has not been thrust or subducted beneath the continents but the continents themselves have moved (Fig. 5-5). A good example is the Atlantic Ocean and the coasts of South and North America on one side and Africa and Europe on the other side. A major result is that the continental margins, and especially the continental rises, in this region have been exposed and received sediments for at least the past 200 million years. Because of this, the sediment sequence is very thick and has considerable oil-bearing potential. In general most of the continental rises of the world occur within plates instead of on their edges, where they would be destroyed by being thrust under the continents (compare left- and right-hand sides of Fig. 5-5c).

Oil, Gas and Sulfur

Oil, gas, and sulfur are easily the most valuable marine resources. The present yearly value of marine oil and gas exceeds $40 billion, which is more than all marine biologic resources. With increased interest in developing energy sources, there will be even further exploration for oil and gas in the ocean (Figs. 5-6 and 5-7). Because of conflicting information concerning petroleum occurrence, its resource potential, how it forms, and other aspects of this resource, it is appropriate to discuss some of the basic aspects of petroleum before its marine occurrence is considered.

Figure 5-6.   An offshore drilling rig. The columns are 10 ft (3 m) in diameter and the working platform is 168 ft (51.2 m) long and 86 ft (26.2 m) wide, with a molded depth of 20 ft (6.1 m) on the sides. The derrick stands 147 ft (44.8 m) high and is on a 30-ft (9.1-m) square base. (Photograph courtesy of Bethlehem Steel Corporation)

Figure 5-7. The self-propelled, semisubmersible drilling rig SS-3000, a joint engineering effort of Bethlehem Steel and Zapata Marine Drilling. This "new generation semisubmersible" will be able to withstand hurricane force winds and 100-ft (30.5-m) waves. The rig design provides for future incorporation of dynamic positioning equipment for operations in water depths beyond the continental shelf. Cost of the rig, including nonshipyard services and equipment, is estimated in excess of $23 million. The rig will be 350 ft (106.7 m) long and 210 ft (64 m) wide and will measure 125 ft (38.1 m) from the keel to the underside of the upper platform. (Courtesy of Bethlehem Steel Corporation)

Origin of Oil

The world "petroleum" comes from the Greek words for oil and rock. The American Geological Institute has defined petroleum as

> a material occurring naturally in the earth composed predominantly of mixtures of chemical compounds of carbon and hydrogen with or without other non-metallic elements such as sulfur, oxygen, nitrogen, etc. Petroleum may contain, or be composed of, such compounds in the gaseous, liquid, and/or solid state, depending on the nature of these compounds and existing conditions of temperature and pressure. (American Geological Institute 1953).

Compounds of carbon and nitrogen only, called hydrocarbons, generally make up over 90% of crude oils (oil is petroleum in the liquid state). The hydrocarbons can vary considerably as to their molecular type and size. The number of the different compounds within an oil is considerable and no crude oil has ever been completely analyzed. Oils having relatively low contents of sulfur are generally more valuable because they create less noxious gases, etc., when burned.

Petroleum originates from the organic remains of organisms, mainly plants that once have lived in the sea or in rivers and that after death have settled to the bottom. The chemistry of the conversion process whereby the organic material changes into petroleum is extremely complex and not completely understood. It is known, however, that if the organic material is oxidized, it does not form petroleum. Organic material can be preserved and accumulate if it settles into an oxygen-poor environment, or if the sedimentation rate is sufficiently high to bury the material before it is completely oxidized. It has been estimated that over 95% of the world's oil initially came from sediments that were deposited in a marine environment, although some of the organic material may have been of terrestrial origin and carried in by rivers. When the organic material reaches the bottom, the conversion process starts, first by bacteria or by other organisms that digest the material and redeposit it as fecal material. As the material is buried and incorporated into the sediments, additional chemical changes occur, in large part caused by the heat and pressures associated with burial. Petroleum does not form if the sediments are not sufficiently buried or if the organic material has been heated too high. After organic material is preserved, under the right but poorly understood chemical conditions, it can eventually, after a long period of time, evolve into a hydrocarbon deposit. The duration needed for the formation of a hydrocarbon deposit is unknown, but such deposits are rarely found in rocks that are less than 2 or 3 million years old; however, they can be found in rocks that are as old as several hundreds of millions of years. Petroleum has been found in near surface fields and from depths of over 6000 m.

As organic material is converted into petroleum, the first compounds formed are those with relatively high molecular weights; these are the viscous or heavy oils. With further conversion and higher temperatures the large molecules are broken into smaller, lighter, and more easily moving molecules. Initially, the organic material is often found with fine-grained clay or mud sediments that have been deposited under reducing conditions. These sediments eventually become rocks, commonly called source rocks, but petroleum is rarely recovered from these source rocks. While the petroleum is forming or after its formation, it usually migrates to more coarse-grained rocks, such as sandstones or reef deposits, which have larger pore spaces and through which the oil can flow (i.e., the rocks have high porosity and permeability). These rocks are commonly called reservoir rocks, and their presence is necessary for an economic deposit. The permeability and porosity of the fine-grained source rocks is usually such that any oil that has accumulated within these rocks is difficult to recover. The migration of the petroleum from fine-grained source rocks to coarser grained reservoir rocks is slow and generally results from the squeezing out of interstitial water and petroleum from the fine-grained sediments. The water and oil will move in the direction of least resistance and can travel either laterally or vertically. However, when an impermeable layer intersects the migrating fluid, the oil or gas can be trapped (Fig. 5-8) and recovered.

In summary, the key conditions for the formation of an oil or gas deposit are:

1. Initial deposition and preservation of organic material, generally in a marine basin

2. Burial by sediments and, under appropriate but not completely understood conditions, transformation of the organic material into petroleum

3. Migration of the petroleum from fine-grained source beds to coarse-grained reservoir beds

4. Geologic traps that restrict the movement of the oil or gas and concentrate it.

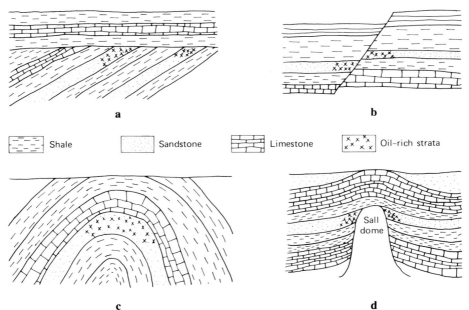

Figure 5-8, a–d. Different possible types of oil traps. For simplicity, the reservoir beds are always indicated as sands; a stratigraphic trap; b structural trap; c anticline; d salt dome or diapir.

The concentration of petroleum in sediments can be considerable, reaching as much as 1000 barrels or more per acre-foot in some reservoir rocks, whereas in nonreservoir rocks it is generally more dispersed, being on the order of 1–10 barrels per acre-foot. Actually, petroleum is very common and occurs almost everywhere in marine sediments. The problem is finding it concentrated in sufficient amounts within a reservoir rock to make it commercially valuable.

Searching for Oil

In searching for an oil or gas deposit, regions having a thick sequence of marine sediments with occasional coarse sediment layers (reservoir beds) are usually considered as having high potential. It must be emphasized, however, that drilling is always necessary to determine whether oil or gas is actually present. Once a potential oil field is found it may take several years to obtain production. Usually 2–5 years are needed to complete the initial survey and to finish sufficient exploratory drilling before the quantity and quality of the deposit is ascertained. Generally, another 2–3 years are needed for production drilling and storage and transportation facilities to be established.

There are some technical limitations to drilling in the marine environment. Most drilling platforms are either fixed or semifixed (the latter can move after drilling) to the sea floor (Fig. 5-6) and so are limited to drilling in depths of generally less than 200 m. A floating platform can essentially drill in any depth of water and maintain its position over the drill site by dynamic positioning on subsurface buoys (see Fig. 5-10). Exploratory wells have already been drilled in water depths of over 1300 m but production has been limited to depths of about 700 m. Drilling in the ocean is considerably more expensive than it is on land; the costs for a well in 200 m of water can easily exceed several million dollars. A major problem with drilling in deep water concerns reentry; this results from the need to replace the drilling bit after it wears out by removing the entire drill string, replacing the bit, and then finding the original hole (reentry) and continuing the drilling operation. If the original hole cannot be found, a new hole must be started. The solution of the reentry problem and the development of dynamic positioning techniques in deep waters have been aided considerably by the work of the drilling ship *Glomar Challenger* (Figs. 5-9 and 5-10). This vessel, which does scientific drilling (not looking for oil) throughout the oceans of the world, has already drilled in water depths of over 6000 m.

Another problem for deep-water drilling is the production operation that must be implemented if gas or oil is found in sufficient quantities. After a successful well is drilled it is necessary to attach, on the sea floor, a complex series of valves and pumps to control the flow of oil or gas. This device, commonly called a "christmas tree," generally has to have some human involvement in its installation and for its subsequent servicing. This can be

Figure 5-9. The deep-sea research drilling vessel, *Glomar Challenger*, which is drilling and coring for ocean sediment in all the oceans of the world. *Glomar Challenger* weighs 10 400 tons and is 400 ft (122 m) long, and the million-pound hook-load capacity drilling derrick stands 194 ft (59 m) above the waterline. She is the first of a generation of heavy drilling ships capable of conducting drilling operations in the open ocean, using dynamic positioning to maintain position over the bore hole. A reentry capability has been established that enables the changing of drill bits and reentering the same bore hole in the deep ocean. Forward is the automatic pipe racker, designed by Global Marine, Inc., which holds 24 000 ft (7315 m) of 5-in. (12.7-cm) drill pipe. (Photograph courtesy of the Deep Sea Drilling Project, Scripps Institution of Oceanography)

done by divers in shallow water, divers working out of submersibles for moderate depths, and possibly by submersibles in deep water. The completion problems associated with getting a well ready for production have limited the water depths for such work. Recently, divers working under controlled laboratory conditions have been able to operate at depths close to 700 m, although working at such depths in the real ocean may never be achieved. Further developments of oil and gas from the deep water will probably need new production systems that can be safely controlled from the surface.

Use of Hydrocarbons

The rate of consumption of hydrocarbons has generally been increasing throughout the world. It is estimated that by 1990 the free world may use as much as 100 million barrels of oil (1 barrel = 42 gallons) a day. In 1978 average production was close to 60 million barrels of oil per day. There have been numerous estimates made of oil reserves and undiscovered poten-

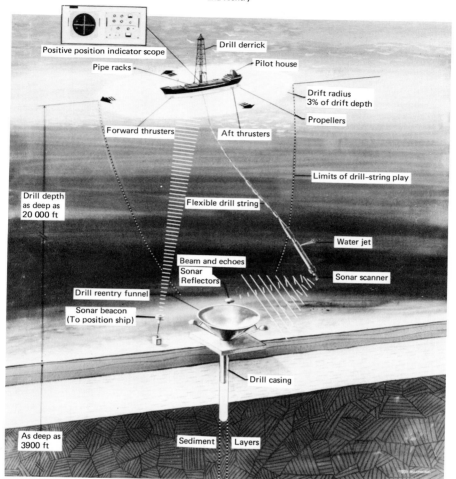

Figure 5-10. The drilling vessel *Glomar Challenger* uses dynamic positioning to hold its station above a sonar sound source placed on the ocean bottom while drilling. Two tunnel thrusters forward and two thrusters aft, along with the vessel's two main propellors, are computer controlled to hold position without anchors in water depths up to 20 000 ft (6096 m) so that drilling and coring can be accomplished. When a drill bit is worn out, it is now possible to retract the drill string, change the bit, and return to the same bore hole through a reentry funnel placed on the ocean floor. High-resolution scanning sonar is used to locate the funnel and to guide the drill string over it. Operational reentry was first achieved on Christmas Day 1970, during drilling in the Caribbean Sea. The artist's concept shows a sonar beacon used for dynamic positioning and a sonar scanner at the end of the drill string searching for the three sonar reflectors on the reentry cone. The relative position of bit and funnel are displayed at the surface on a positive position indicator scope. The Deep Sea Drilling Project developed reentry when stopped short of scientific goals at many bore holes in the Atlantic and Pacific Oceans when the bit hit beds of chert or flintlike rocks that dulled the bit and forced early abandonment of bore holes. (Photo courtesy of the Deep Sea Drilling Project, Scripps Institution of Oceanography)

tial, etc., and it is beyond the scope of this book to discuss the criteria or validity of the calculations. Nervertheless, regardless of whose data are used or what assumptions are made, one point is fairly obvious: The supply of oil and gas cannot keep pace with present and future rates of consumption. There are other alternatives, such an nuclear power, coal, solar energy (see Chapt. 9), mining of oil shale (a sedimentary rock that yields oil when crushed and distilled), and the like that could reduce our future dependence on petroleum, but the prospects for these replacing petroleum completely do not look too good under today's economic, political, and social restrictions. A breakthrough in technology or an increase in market prices may make some submarginal deposits become real reserves. For example, cumulative oil production in the United States has been almost 100×10^9 barrels, but as many as 400×10^9 barrels have been left behind in the ground because of the high cost or lack of technology necessary to remove it. A technology that could increase recovery by just 1% over the present average of 30%, to 31% of the total oil, at an acceptable cost, could add 4×10^9 barrels, or close to a year's supply, to United States proved reserves.

Offshore Oil—The United States: A Case Study

A discussion of offshore oil potential for the countries of the world is beyond the scope of this book. Instead, a brief description of United States' oil and gas potential is given.

The total area of the United States' continental shelf and slope to a water depth of 8000 ft (2438 m) is about equal to the United States' land area and it might be assumed that the area will yield about the same amount of oil as has already been produced from land, i.e., about 100×10^9 barrels of oil. Some have suggested that the number could be twice as high. Weeks (1968) estimated that the continental shelf and deeper areas out to a depth of 1000 ft (304 m) could contain 140 billion barrels of oil, plus 60 billion barrels of petroleum that could be obtained by secondary recovery techniques. In 1972, the National Petroleum Council projected potentially recoverable United States oil and gas at 386 billion barrels and 1178 trillion cubic feet, half of which could come from Alaska and offshore areas. In early 1974 the U.S. Geological Survey estimated that between 200 and 400 billion barrels of oil and between 1000 and 2000 trillion cubic feet of natural gas can be found. These numbers have been challenged by some representatives of the oil companies, especially Mobil, as being too high (Table 5-3). If Mobil's estimates are correct, less than a 20 year supply remains to be discovered, and only about 12–15 years of reserve are presently available. The U.S. Geological Survey's estimate was based mainly on extrapolation of past findings to future possibilities; i.e., if only 20% of the total area had been explored, the remaining 80% should yield four times what was found in the original 20%. Oil companies argue that the prime areas were explored first and that the remaining areas should contain proportionately less oil. Interestingly,

Table 5-3. Difference in Estimates of United States Undiscovered Recoverable Oil
and Gas Resources

Location	Undiscovered Recoverable Oil and Natural Gas Liquids (billions of barrels)			Undiscovered Recoverable Natural Gas (trillions of cubic feet)		
	Mobil expected value[b]	USGS Low	High	Mobil expected value[b]	USGS Low	High
Onshore						
Alaska	21	25	50	104	105	210
Lower 48 states	13	110	220	65	500	1000
Subtotal onshore	34	135	270	169	605	1210
Offshore						
Atlantic	6	10	20	31	55	110
Alaska	20	30	60	105	170	340
Gulf of Mexico	14	20	40	69	160	320
Pacific coast	14	5	10	69	10	20
Subtotal offshore	54	64	130	274	395	790
Total United States	88	200	400	443	1000	2000

[a]Adapted from *Science*, 1974, 185, pp. 127–130.

[b]Mobil estimates include water depths to 6000 ft (1829 m), whereas USGS now stops at 600
ft (183 m). Mobil's numbers represent the median value of a probability distribution. For instance, there is a 90% chance that total United States oil is greater than 50 billion barrels and
less than 150 billion; the expected value is 88 billion.

the oil company estimate suggests more oil resources in the offshore environment than on land.

Other areas of oil and gas potential, besides the continental shelf, are the
deeper parts of the continental margin: the continental slopes and especially
the continental rise. The continental slope probably is not as favorable for
petroleum as the shelf landward of it or the continental rise seaward of it.
This is mainly because of the relatively slow rate of sedimentation on the
slope, which permits much of the organic material to be oxidized, and
because of the general absence of reservoir beds.

In 1975 total United States oil and gas production was 3056 million barrels of oil and 20 100 billion cubic feet of gas. Of this total, production from
the outer continental shelf (the area beyond 3 miles, or about 5.5 km, from
the coast) was 330 million barrels of oil (10.8% of total) and 3458 billion
cubic feet of gas (21.2%). Of this offshore total, 21% of the oil and 90% of
the gas came from Louisiana waters.

The amount of money that the United States Government has obtained
from offshore lease sales has been immense. Over $14 billion has been
received from October 1954 to December 1975 (Table 5-4). Most of these

Table 5-4. Summary of United States Outer Continental Shelf Leasing from 13 October 1954 to 31 December 1975[a]

State	Lease Area (1000 acres)	Bonus (million dollars)	Percent of Total		Average Cost ($/acre)
			Area	Bonus	
Alabama	74	136	0.6	0.9	1 838
California	988	1 054	7.9	6.6	1 067
Florida	490	1 102	3.9	6.9	2 249
Louisiana	7 344	9 402	58.4	59.1	1 280
Mississippi	35	116	0.3	0.7	3 314
Oregon	425	28	3.4	0.2	66
Texas	3 053	4 073	24.3	25.6	1 334
Washington	155	8	1.2	0.1	52
Total United States	12 565	15 917	100	100	1 267
Total Gulf of Mexico	10 997	14 827	87.5	93.1	1 348

[a]Data from *Ocean Industry*, May 1977, p. 75. Totals are rounded off.

lease sales also require a royalty payment of $16\frac{2}{3}\%$ (actually a relatively small royalty) of the value of material produced and some smaller fees. The money obtained from lease sales goes directly to the Treasury Department and is not used to reduce the cost of oil or to aid in obtaining it. In recent years lease sales have brought even higher sales; for example, the first lease sales in the Atlantic brought in $1.1 billion in 1976 for leases in the Baltimore Canyon area off New Jersey and Delaware.

Offshore Oil—Other Areas

In 1973 the total world oil output was estimated at about 55.5 million barrels per day and about 10 million barrels, or about 18% of the total, came from offshore fields. By 1976 total production reached 57.3 million barrels per day but only about 8.6 million barrels, or about 15%, came from offshore. This drop was because some Middle East offshore fields were operated at less than full capacity. Offshore areas are estimated to hold about 23% of the world's crude oil reserves and 14% of the gas reserves (*Oil and Gas Journal*, August 1977). Also of importance is that less than 20% of offshore reserves have been produced, whereas at least 43% of conventionally recoverable land resources have been used. Most of the offshore oil already found is in inland seas, such as the Gulf of Mexico and Persian Gulf, or on shallow parts of the shelf (many times a continuation of a land deposit). Over 40% of the new major oil fields (500 million barrels or more) in recent years have been found in the offshore environment. For example, some rather large discoveries have been recently made off the east coast of Mexico.

The ultimate amount of oil obtainable from the marine environment could be considerable, when compared to that left to be found on land. This in part is because most land areas have been explored and promising areas already drilled, whereas exploration in the sea has been restricted to relatively shallow waters. Actually, exploration in the marine environment has been fairly conservative and most production occurs adjacent to productive areas on land. Exploration, however, of some sort has already taken place off the coasts of most countries that have marine boundaries.

An offshore area especially favorable for oil or gas accumulation is river deltas. These regions generally have basins containing thick sequences of sediments that extend from land into the offshore area (Fig. 5-11). In addition, rivers generally carry large amounts of nutrients. Offshore organic production therefore is usually high—a point also favorable for hydrocarbons.

Marginal seas (adjacent to or surrounded by continents), such as the Black Sea, the Mediterranean Sea, or the East China Sea, are also favorable sites for petroleum accumulation. Seismic reflection studies generally show that these areas contain thick sedimentary sequences with occasional diapirs or other promising structures. Coring of the surface sediment usually shows

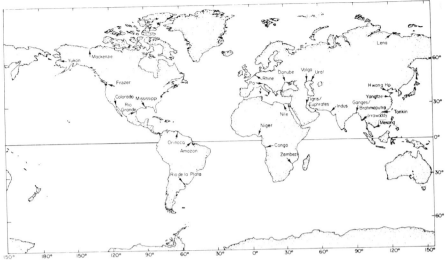

Figure 5-11.   Locations of the major river deltas in the world.

relatively high amounts of organic material that, if present at depth, indicates conditions favorable for hydrocarbons.

A major possible future source of petroleum may be the continental rise (Fig. 5-12), an area usually containing a thick wedge of sediment situated at the base of the continental slope and extending seaward. The distribution of continental rises is strongly controlled by sea-floor spreading. Where oceanic plates are being thrust under the continents, rises generally do not form because the sediment is removed by the subduction process. Rises, in contrast, are present along continental margins where subduction does not occur (Fig. 5-13). Continental rises and partially filled sedimentary basins cover an area of about 19 million square kilometers (Menard and Smith 1966) which is equal to about 33% of the total continental shelf and slope area.

Continental rise deposits generally start at depths of 3000 m or more and contain sequences of sediments that may include coarse-grained turbidities (reservoir beds) within normal finer grained pelagic sediments. Much of the sediment forming the rise has slumped down from shallower areas and has carried with it relatively high amounts of organic material. Thick sequences of sediments in deep water are also common off large rivers, such as the Amazon, Indus, Mississippi, Congo, and Nile. Geophysical studies off some of these rivers and on continental rises have occasionally revealed diapir or salt-dome structures resulting from the upward movement of a "blob" of less dense salt or mud through the overlying sediment (Fig. 5-8d). This upward moving material bends the sedimentary layers, forming structures that can act as traps for oil and gas.

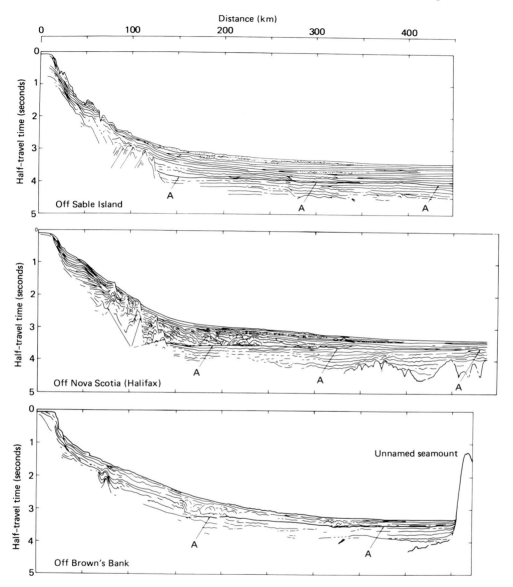

Figure 5-12. A series of seismic profiles made across the continental shelf, slope (left-hand side of figure), and continental rise of the east coast of the United States. These and other such records generally show thick sequences of sediment on the continental rise (the scale on the left is equivalent to about 2-km thickness for each second of half-travel time). Reflector A refers to a distinct acoustic reflector that can be correlated from one profile to another and that has been drilled by the *Glomar Challenger* and found to be a hard chert layer. (Adapted from Emery and others 1970)

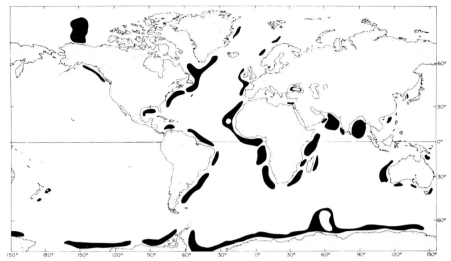

Figure 5-13.   Distribution of continental rises (blacked-in areas).   (Adapted from Emery 1969, and others)

Sulfur may often be found as part of the cap rock of salt domes that themselves are buried beneath the sediments of the continental shelf or continental rise.   The salt, probably deposited when the basin was under evaporitic conditions, can flow through overlying heavier sediments moving upward as a long, thin plug producing a domelike structure; hence the name salt dome. These features are extremely common in the Gulf of Mexico and have formed traps for some of the oil recovered from that area.   As the salt nears the surface, some of it is dissolved by ground water and a residue or cap rock of relatively insoluble anhydrite remains.   Anhydrite ($CaSO_4$) can interact with petroleum and bacteria, forming hydrogen sulfide gas, water, and the mineral calcite ($CaCO_3$).   If oxygen then reacts with the hydrogen sulfide, water and elemental sulfur can be produced and concentrated within the cap rock.   The sulfur can be recovered fairly simply by the so-called Frasch process, where superheated water and air are piped down the drill hole. This melts the sulfur and allows the sulfur and water to be pumped up to the surface.

Salt domes off Louisiana have already yielded 200 million tons of sulfur. However, fewer than 10% of the drilled salt domes have commercial amounts of sulfur.   The discovery by *Glomar Challenger* of sulfur-bearing domes in the deep parts of the Gulf of Mexico has encouraged an increase in the search for offshore sulfur deposits.

Emery (1974b) has detected features in sediments off the west coast of Africa that may be caused by gas hydrates or clathrates.   They were found in the upper 50 m of sediment and he has called them pagoda structures (Fig. 5-14).   He suggests that these structures result from cemented sediment, where the cementing agent is the gas, methane. Methane would behave like ice at the pressure and temperature found on the ocean floor. Emery (personal communication) nevertheless recommends caution about considering this a resource: feeling that even if methane is associated with pagoda structures, there is considerable doubt that it would have commercial value in the near future.

It seems very unlikely, however, that the deep abyssal plains, beyond the continental rise, and other parts of the ocean have economically valuable hydrocarbon deposits. These regions are unfavorable for several important reasons, including the lack of sufficient thickness of the sedimentary rocks; the low sedimentation rate, which permits, among other things, the organic matter to be oxidized; and the general absence of reservoir rocks.

In their normal research in the ocean, marine geologists and geophysicists can often locate areas of oil and gas potential, simply by detecting thick sedimentary sequences and turbidite layers from seismic profiler records. This

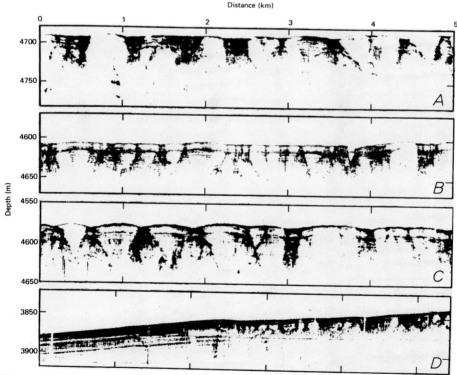

Figure 5-14. Pagoda structures on abyssal plains and lower continental rises off western Africa. Whitish triangles (sections of cones) are believed to have been cemented by methane hydrates, or clathrates. Seismic reflection profiles show basement–sediment interfaces at roughly 4700 m (A), 4600 m (B) and (C), and 3850 m (D). (From Emery 1974b)

type of work has caused some countries to look suspiciously at such research and some of the more vociferous have voiced fears of exploitation (see Chapt. 3). Actually, the delineation of a petroleum deposit requires much more detailed exploration than that generally ever done by a research scientist. Nevertheless, the basic research done by a marine geologist or geophysicist could provide the basis for discovery of many types of mineral deposits. However, in any situation drilling still must be done before a hydrocarbon deposit can be verified.

The scope and scale of future development and search for oil in the ocean will be affected by land oil activity and by development of other sources of energy. Oil shale and tar sand deposits on land, in particular, may also become new sources of oil. It has been estimated that one such deposit, the Green River Formation in the western United States, contains as much as 2 trillion barrels of oil, of which 1.3 trillion barrels may be fairly easily exploited. The problem is threefold in that the cost should be competitive with other sources of energy, there should be acceptable or no environmental damage, and the necessary technology and leadership must be available for research and development. The 1973 Arab embargo and subsequent fourfold increase in the price of oil may have helped the cost of oil shale to be competitive with foreign prices. The environmental damage, however, can be considerable as the mining (strip mining, mainly) and refining operation produces considerable waste and consumes large amounts of water. The development of technology has not proceeded sufficiently far enough for any large-scale operation in the United States, although the USSR has claimed production of over 160 million barrels of oil from oil shale. If oil shale can be produced at a compatible cost and with more environmental safety than offshore oil (which can be done fairly safely—see Chapt. 6), then further exploration into the deeper parts of the ocean may proceed fairly slowly. The most likely view, however, is that offshore oil and gas will be a major industry and source of energy for decades to come.

Resources of the Deep Sea

The deep sea, which includes the main ocean basins, trenches, rises, and the oceanic ridges, totals about 79.4% of the ocean and covers an area of about 287 million square kilometers or about 110 million square miles. In size the deep sea is about four times the size of the continental margin. The mineral resource potential of the deep sea, however, appears to be considerably less than that of the continental margin. The major resource of the deep sea is probably managanese nodules; of less potential are deep sea muds and oozes and deposits formed on the oceanic ridges.

Some recent scientific discoveries may lead to additional resources in the deep sea. The implications of the sea-floor spreading and plate tectonics concept and how it has affected the economic potential of the deep sea are an example. Recent discoveries of heavy metal accumulations along zones of active spreading and zones of subduction could possibly become economically significant as further studies are made, but the probabilities are small.

Deep-Sea Muds and Oozes

Much of the deep ocean floor is covered by fine-grained mud deposits. They are generally classified according to their most common constituent. The most common deposit is called, simply enough, brown clay and it is present almost everywhere in the deep ocean, although it frequently is diluted with other sediment types, including manganese nodules. Brown clay, also sometimes called red clay, has a very slow accumulation rate on the order of a few millimeters per thousand years. Areas of brown clay deposition rarely have more than a few hundred meters of total sediment thickness, in large part because of the slow sedimentation rate and in part because of sea-floor spreading movements of the ocean floor. These movements can cause these sediments either to be subducted under (Fig. 5-5) or to be accreted onto a continental plate; alternatively they can be buried by continental rise sediments.

Brown clay deposits cover an area of about 40 million square miles or about 100 million square kilometers and, assuming an average thickness of 300 m, they have a volume of about 30 million cubic kilometers. Chemical analyses of the brown clays show as much as 9% aluminum, 6% iron, and smaller amounts of copper, nickel, cobalt, and titanium. Some of these metals are more enriched in the brown clay than in rocks mined on land and it is therefore logical that brown clays eventually may also be mined. Mero (1965) has estimated that brown clays contain enough aluminum and copper that, if they could be economically mined, there would be a supply that would last over 1 million years at present rate of consumption. The "if" is a big one and among the major problems are recovery from depths of 6000 m or more and refining the fine-grained material. One advantage of the deposit is that it lies unconsolidated directly on the sea floor with no overburden.

Muds that contain relatively large amounts of shells or tests (usually more than 30%) of dead organisms are called oozes. These oozes are of two main types: calcareous oozes or siliceous oozes, depending on whether they are composed mainly of shells made of calcium carbonate or of shells made of siliceous material. They can be further subdivided according to the major shell type, such as into foraminiferal ooze, diatom ooze, radiolarian ooze, etc. Collectively, these oozes exceed brown clays in areal extent, although the two are not mutually exclusive; it is a question of which dominates, and this is generally controlled by near-surface oceanographic conditions that influence the growth of the organisms and the depth of water, because calcium carbonate tends to dissolve below around 5000 m. The deposition rate of oozes is of the order of a few centimeters per thousand years, which, although extremely small, is about 10 times the rate of deposition of the brown clays. Under areas of high biologic activity, therefore, the settling shells will generally dilute any brown clay deposit.

Calcareous oozes may cover as much as 50% of the ocean floor and can have carbonate contents as high as 95%, a content high enough that they could be used as a source of limestone for cement. As in the case of brown clays, the volume of this material is awesome and, if ever mined, could supply limestone at a rate equal to several million years of consumption. Siliceous oozes could be mined for their silica content, which can be used for insulation and soil conditioners. This deposit also covers an extensive area of the sea floor, although somewhat less than that of calcareous ooze.

Another potential clay resource from the deep sea is zeolite, a mineral of which phillipsite is the most common form. This mineral apparently forms on the ocean floor from the decomposition of volcanic glass. Phillipsite is extremely common in the Pacific and may be one of the most abundant minerals on the earth's sufface; it could be mined for potash and used for fertilizer.

All the oozes and muds mentioned above must be considered only as potential resources based on present-day needs and the technologic difficulties of mining them. Even so, numerous people have been impressed by their vast extent and have suggested that they could become very important potential resources in the near future. There are some aspects of this that are especially appealing. One is that although the rate of accumulation is amazingly small, the accumulations extend over such a large area that the net rate of accumulation of several elements is considerably higher than the present rate of their consumption on land. One type of deep-sea mud, the so-called Red Sea metalliferous muds, may be mined before the decade is over.

Red Sea Metalliferous Muds

An especially interesting potential mineral resource of the deep-sea floor was discovered by accident in the Red Sea. The deposit was found in an area of active sea-floor spreading in the central part of the Red Sea, where Africa and the Saudi Arabian Peninsula are slowly moving away from each other. In 1948 the Swedish research vessel *Albatross* made a routine hydrocast in this area and noted slightly anomalous values for temperature and salinity of water collected along the bottom of the central ridge of the Red Sea. At the time they assumed that the anomalies were caused by instrument error. Further study in the area was done from the British R.R.V. *Discovery* and ships from the Woods Hole Oceanographic Institution, and these finally led to the discovery in the early 1960s of three pools of hot saline water at a depth of about 2000 m on the bottom of the Red Sea (Swallow and Crease 1965, Miller and others 1966, Degens and Ross 1969). Salinity of these strange waters was as high as $257°/\infty$[2] and temperatures were then as high as 56°C (138°F).[3]

---

[2]Not salinity in the usual sense of the word, because the ratios of the major elements in the Red Sea are not similar to the ratios found in normal sea water.

[3]By 1977 some waters had heated up to over 60°C, or 140°F.

Figure 5-15.   The hot brine area of the Red Sea. Hot salty water underlain by sediments enriched in several heavy metals has been found in the cross-hatched areas (contours in meters). (From Ross and others 1969)

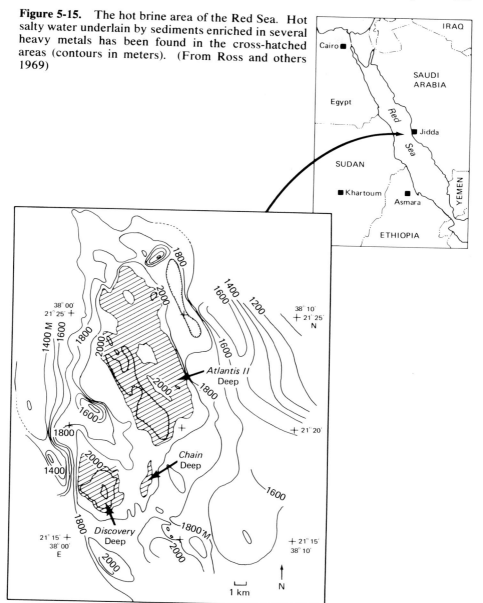

During a major expedition in 1966 to the Red Sea by the Woods Hole Oceanographic Institution, the area was mapped (Fig. 5-15) and numerous sediment samples were taken. The economic value of these deposits was first noted by Dr. Frank T. Manheim who estimated that the top 10 m of the sediment underlying one of these pools, the Atlantis II Deep, had an *in situ*

Table 5-5.   Gross Value of Metals in Upper 10 Meters of Sediments Collected from Atlantis II Deep Based on 1967 Metal Prices[a]

Metal	Average Assay (%)	Tons	Value ($)
Zinc	3.4	2 900 000	860 000 000
Copper	1.3	1 060 000	1 270 000 000
Lead	0.1	80 000	20 000 000
Silver	0.0054	4 500	280 000 000
Gold	0.0000005	45	50 000 000
Total			$2 480 000 000

[a]Adapted from Bischoff and Manheim (1969).

value of about $2 billion. A later, more detailed report gave a value of $2.4 billion for the same material (Table 5-5). This value does not consider the cost of raising the sediments from 2000 m and refining and marketing the minerals but is only their value on the sea floor.

In February 1968, a San Francisco company, Crawford Marine Specialists, Inc., applied to the United Nations for an exclusive mineral exploration lease to the area, but it was not granted because the United Nations claimed that they had no right to grant it. At about the same time representatives of a Los Angeles company, International Geomarine Corporation, after several visits to Sudan, were able to obtain a lease from that government. It was then thought that the hot brine deposits lay more toward the Sudanese than the Saudi Arabian side of the Red Sea. This was because the earlier studies did not use sophisticated navigation techniques and the actual location of the area was in error; later studies showed that the heavy metal deposit was almost equidistant from the Sudan and Saudi Arabian coasts. Based on the general principles of international law, the resources of the Red Sea would belong to the countries bordering the sea, and ownership would be decided by where the deposit occurred relative to a median line equally dividing the Red Sea. Other companies claimed the brine deposits, including a group from Ethiopia and another that was incorporated in Leichtenstein. This was followed by a claim by the Saudi Arabian government that they owned the mineral resources of the entire Red Sea but would be willing to negotiate such a claim. Recently Saudia Arabia and Sudan have formed a joint commission to mine and share the resources of the area. They have contracted with a German mining company, Preussag AG, and mining may begin in the early 1980s.

Although several attempts were made to locate new brine pools, it was not until the results of a series of detailed German studies made from the research vessel *Valdivia* were published that it was found that 13 new brine pools had been discovered in the Red Sea (Bäcker and Schoell 1972). The Atlantis II Deep still appears to be the most valuable.

What makes these deposits so interesting is not necessarily their high content of copper, zinc, and silver, but the fact that scientists can actually "see" a mineral deposit forming on the sea floor. Similar deposits are found on land, but they are hundreds of millions of years old. It follows that if such a deposit is actively forming now on the Red Sea floor, in an area of sea-floor spreading, that similar deposits should occur in other parts of the ocean. Before the Red Sea impact was fully known, Bostrom and Peterson (1966) had noted iron enrichment in sediments callected from the East Pacific Rise, also a zone of sea-floor spreading. The small amount of sampling done so far along the oceanic ridges suggests that hydrothermal and volcanic activity may, in some places, be concentrating metals within the bottom sediments and rocks.   Sediments collected from some localities on the active midoceanic ridges (Mid-Atlantic Ridge, East Pacific Rise, and Mid-Indian Ocean Ridge) have shown enrichments in cobalt, copper, chromium, iron, manganese, nickel, uranium, and mercury, with lesser amounts of bismuth, cadmium, and vanadium. These concentrations are generally not economic, but occasionally some especially high (but small in extent) concentrations occur. Small veins of pure copper have been recovered by the Deep Sea Drilling Project at two locations, in the Indian Ocean near the equator and some 560 km southeast of New York City in sediment under the lower continental rise in a water depth of about 5000 m. Recent investigations using the submersible *Alvin* have detected hot water coming from a rift area near the Galapagos Island and off Baja, California. Both areas also have a unique fauna consisting of giant worms, clams and other creatures (Fig. 5-16) that are dependent on the hot water for their sustenance. Off Baja the waters, perhaps as high as 400°C in temperature, are presently forming a mineral deposit on the sea floor. None of the above has yet attained the economic potential of the Red Sea, and for various geologic reasons they probably will not be economically important.

Manganese Nodules

Certainly one of the more interesting and probably valuable resources of the deep-sea floor is manganese nodules (sometimes called iron–manganese or ferromanganese deposits). These deposits can occur as round spheres from about 1–20 cm in diameter (Fig. 5-17a), as coatings on rocks and other objects, or as long slabs, commonly called manganese pavements (Figure 5-17b). Regardless of their form, they are mainly composed of hydrated oxides of iron and manganese. They generally form around a nucleus, such as a small shell, a rock, or a shark's tooth, and slowly grow outward in concentric rings. Ferromanganese deposits are found throughout the deep-sea marine environment, as well as in lakes or other shallow-water bodies. They were first discovered during the *Challenger* expedition (1872–1876). The economic interest in the nodules is principally because of their accessory elements—copper, nickel, and cobalt—as well as their main component, manganese, and their supposed ease of recovery from the sea floor.

a

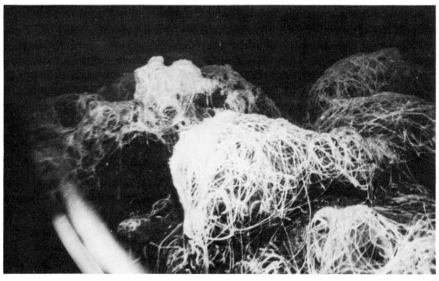

b

Figure 5-16 **(a)** Cluster of giant sea worms found in the hot springs in the Galapagos Rift. The worms have a brilliant red tip and live in a one-inch diameter tube attached to rocks on the ocean bottom. An eight-and-one-half-foot specimen was recovered; the worms are related to worms which normally reach a maximum of about 12 inches in length. Photo by Holger Jannasch, Woods Hole Oceanographic Institution. **(b)** Dubbed "spaghetti" by scientists because they were draped like threads over rocks, these animals were found in the hot springs area in the Galapagos Rift. The animal was identified as enteropneust, a type of worm. Photo by David Karl, University of Hawaii. Both photographs courtesy of Woods hole Oceanographic Institution.

Figure 5-17 (a). Vast areas of manganese nodules on the floor of the southwestern Pacific Ocean. The average size of the nodules is about 4 in. (10 cm) in diameter. Water depth is 5292 m. (Adapted from Heezen and Hollister 1971)

0.3 m

Figure 5-17 (b). Manganese pavement photographed on the Blake Plateau off the east coast of Florida. (Photograph courtesy of Woods Hole Oceanographic Institution)

Manganese nodules are common to all the oceans (Fig. 5-18), although their concentration may be higher in certain areas. For example, they are especially abundant in areas of low sedimentation rates, such as abyssal plains. Nodules form at an extremely slow rate that may be on the order of 1 mm in a million years, and so they would be easily buried where the sedimentation rate is high. The number of nodules on the sea floor can be immense; one of the highest concentrations observed (as determined by bottom photographs) was 100 kg/m^2 or about 300 000 tons per square mile. A mining operation would need about 30 000–75 000 tons per square mile (Mero 1972). Estimates are that as much as 25% of the sea floor is covered by nodules and that over 1.5 trillion (1.5×10^{12}) tons are in the Pacific Ocean alone.

Calculations on the rate of formation of nodules over the entire ocean suggest a production rate of about 10 million tons per year. This statistic is especially interesting in that some elements are being incorporated within the nodules at a rate greater than their present industrial consumption on land. However, this point has little value because the presently known nodules contain many thousands of years of reserves of these elements.

Numerous chemical analyses have been made of manganese nodules (Table 5-6). These data generally show differences between oceans and some general similarities within certain physiographic provinces within an ocean. The distribution of the most economically important elements—manganese, copper, nickel, and cobalt—has been plotted for each ocean by Horn and others (1973) as part of an International Decade of Ocean Exploration (IDOE) study (Fig. 5-19). These studies have shown that a red clay and siliceous ooze area of the north Pacific looks most favorable (especially the siliceous ooze area) for mining because of its relatively high metal content (Table 5-7). These deposits lie north of the equator in a broad band between 6°30′N and 20°N and extending from 110°W to 180°W (Horn and others 1973), covering an area of 800 by 4200 naut. mi. or about 3.4 million square nautical miles (over 11 million square kilometers). They lie on a relatively smooth area of the sea floor, which could simplify recovery. The nodules of the Atlantic and Indian Oceans are generally characterized by having copper, nickel, and cobalt values below those generally considered economic. Certainly all possible deposits have not been discovered; likewise, some deposits of lower grade could become important if they are near a major market or near areas where mineral refinement is possible. In addition, further changes in the importance or need of one of the minerals contained within the nodules could make a marginal deposit an economical one. For example, shortages in cobalt more than doubled its price during a 6-month period in 1978.

A report by Menard and Frazer (1978) indicates that the amount of copper and nickel in nodules may be inversely proportional to the concentration of nodules on the sea floor. Their analysis, although based on a small number of samples, may have considerable importance for mining opera-

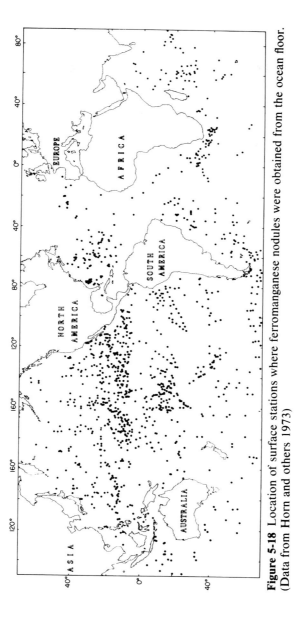

Figure 5-18 Location of surface stations where ferromanganese nodules were obtained from the ocean floor. (Data from Horn and others 1973)

Table 5-6. Average Composition of Manganese Nodules from the Different Oceans

Minerals	South Pacific[a]	North Pacific[a]	West Indian[a]	East Indian[a]	Atlantic Blake Plateau[b]	Atlantic[c]
Manganese	16.61	12.29	13.56	15.83	15	16.18
Iron	13.92	12.00	15.75	11.31	11	21.82
Nickel	0.433	0.422	0.322	0.512	0.4	0.297
Cobalt	0.595	0.144	0.358	0.153	0.3	0.309
Copper	0.185	0.294	0.102	0.330	0.1	0.109
Lead	0.073	0.015	0.061	0.034	—	—
Molybdenum	0.035	0.018	0.029	0.031	0.03	—

[a]Data from Cronan (1967) and Cronan and Tooms (1969) in weight percent, air-dried weight.
[b]Data from Manheim (1972a) in weight percent, air dried.
[c]Data from Cronan (1972) in weight percent, air-dried.

Table 5-7. Average Composition of Manganese Nodules in a North Pacific Area Considered to be Favorable for Mineral Recovery[a]

Mineral	Red Clay Sediments (%)	Siliceous Oozes (%)
Manganese	17.43	22.36
Iron	11.45	8.15
Nickel	0.76	1.16
Copper	0.50	1.02
Cobalt	0.28	0.25

[a]Data from Horn and others (1972b).

tions. Why one area has a higher metal content than another is not completely understood. Some have suggested that the elements are directly removed from the sea water and incorporated into the nodule, which does not explain the areal differences. Others have hypothesized that organisms, such as plankton, concentrate metals within their organic matter and after death are deposited in the bottom sediments. After the decay of the organism the metals are released and can move up to the sediment surface where they may become incorporated within the nodule. G. Arrhenius (in *Science*, 1974), one of the proponents of this hypothesis, notes that sediments in areas of nickel-rich nodules are depleted in nickel, suggesting that this biologic mechanism has occurred. This hypothesis is consistent with finding higher metal contents in nodules from siliceous (of biogenous origin) sediments (Table 5-7).

The potential value of the manganese nodules has excited many about the riches of the ocean and has increased the intensity of the international discussions concerning ownership of the sea floor (see Chapt. 3). Even with this enthusiasm there are many who still doubt the economic value of the nodules. Nevertheless, there have been numerous consortiums and companies formed to mine manganese nodules. One such organization—Summa Corporation (owned by Howard Hughes) and their vessel *Glomar Explorer*—was actually a front for a CIA attempt to raise a sunken Soviet submarine (see Chapt. 7, Fig. 7-4).

Deepsea Ventures, initially part of Tenneco, Inc., a United States company, was probably the first to develop an operational prototype hydraulic dredge system for the recovery of nodules. In 1970, the device was successfully tested in a water depth of about 800 m. A specially designed ship (R/V *Deepsea Miner*, Fig. 5-20), using a 9⅝-in. (24.4-cm) pipe and an air-lift hydraulic suction dredge, was able to lift large amounts of manganese nodules off the sea floor. The success of the prototype system (Fig. 5-21) indicated to the company that recovery from 7000 m is possible. A large ore carrier has been converted (*Deepsea Miner II*) and became operational in

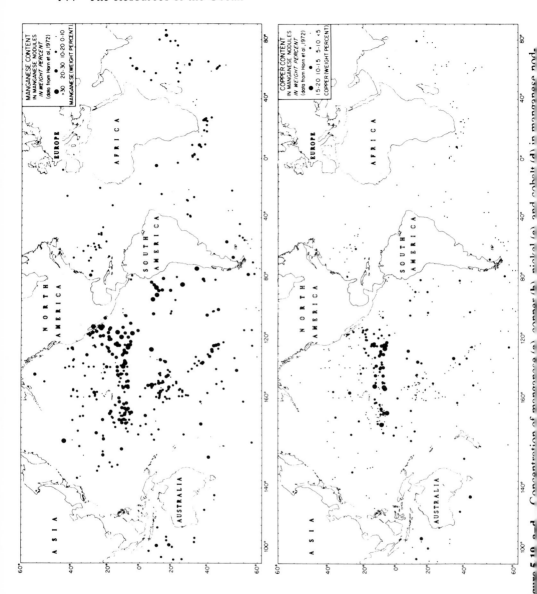

Figure 5.10 a–d. Concentration of manganese (a), copper (b), nickel (c), and cobalt (d) in manganese nod

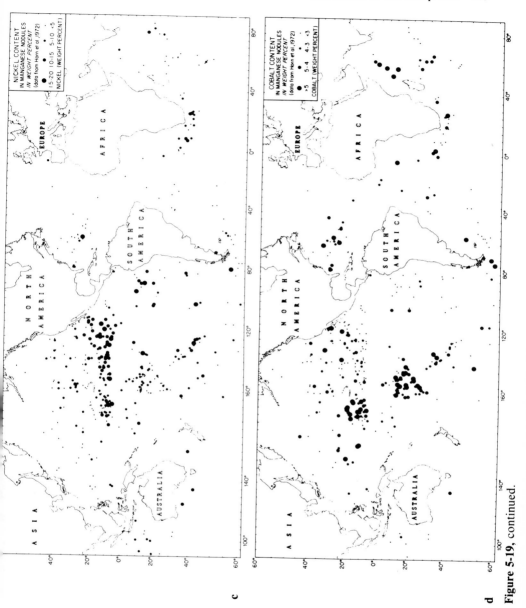

Figure 5-19, continued.

Figure 5-20.   R/V *Deepsea Miner,* a ship built by Deepsea Ventures to mine manganese nodules (see Fig. 5-21) from the ocean floor. (Photograph courtesy of Deepsea Ventures, Inc.)

1977.   Deepsea Ventures has extensive plans to build recovery systems, carriers, and chemical processing plants. In 1970 a Japanese ship, *Chiyadu Maru,* using a continuous line bucket system, succeeded in bringing nodules up to the surface from depths of about 1000 m. Later tests have apparently been successful in deeper water.

Because of the high cost and advanced technology needed to mine the nodules, many of the players in these operations have combined forces and formed consortiums. In January 1974 the United States Kennecott Copper Corporation announced a 5-year, $50 million research and development program to determine the feasibility of mining manganese nodules from the sea floor. The project was to be a financial consortium, with Kennecott supplying 50%, Rio-Tinto-Zinc Corporation (England) 20%, and 10% each from Consolidated Gold Fields Ltd. (England), Mitsubishi Corporation (Japan), and Noranda Mines Ltd. (Canada). (The composition of the various consortiums has changed with time—these are the principal ones as of early 1979.) In 1974 a group called Ocean Mining Associates was formed, consisting of 40% ownership by US Steel, 40% Union Miniere of Belgium, and 20% Tenneco; Deepsea Ventures is the operating arm of this venture. Another manganese consortium is Ocean Management Inc., which includes International Nickel, Inc. (United States), International Nickel,

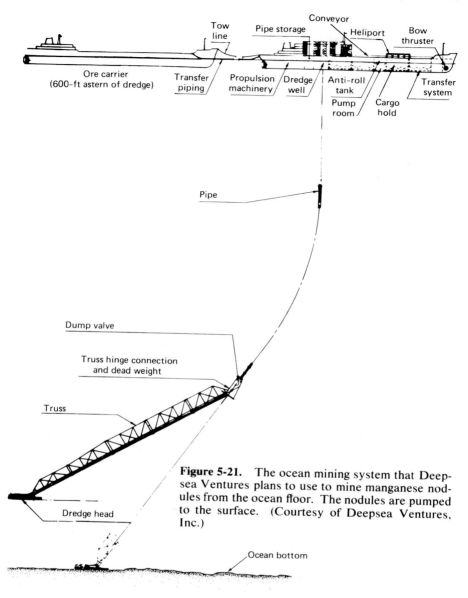

Figure 5-21. The ocean mining system that Deepsea Ventures plans to use to mine manganese nodules from the ocean floor. The nodules are pumped to the surface. (Courtesy of Deepsea Ventures, Inc.)

Ltd. (Canada), AMR Group (several German companies, including Metallgesellschaft and Preussag), 23 Japanese companies, and Sedco (United States). This latter group, which combines expertise in mining, exploration, processing, and marketing, has converted a drilling ship into a mining vessel. They will pump the nodules up using an air-lift system through a pipe system that will be as much as 6000 m long. A relatively new group, Global Marine Development, Inc., is composed of Lockheed Missile and Space Co. (United States), Billiton International Metals B.V. (of Royal Dutch Shell), B.K.W. Ocean Minerals B.V. (another Dutch company), and Amoco Minerals Co. (a division of Standard Oil of Indiana). This consortium has leased *Glomar Explorer* (Fig. 7-4), the vessel built by Howard Hughes, for testing a system to raise manganese nodules. The irony is that the original "cover story" is now becoming a fact.

Deepsea Ventures did a rather interesting thing in 1974 in that it filed a claim for the manganese nodules within a 60000-km² area in the Pacific Ocean. The area lies within the high seas (Fig. 3-1) outside the present jurisdiction of any country. The claim was made to the U.S. Department of State, to several other countries, and to the world at large via a press release. The U.S. Department of State, at that time, said that it did not grant or recognize exclusive mining rights to the mineral resources of an area of the sea bed beyond the limits of national jurisdiction. They also suggested that the Law of the Sea Conference was the place where such international questions should be considered. In later years several bills have been submitted to the United States Congress concerning the mining of manganese nodules by United States companies and these are described in a later section. It should be emphasized that the United States has been the prime mover in trying to develop and mine the manganese nodules in the deep sea. It is also the United States that presently has most of the technology and capital to initiate such a project.

Basically, the problem concerning manganese nodules can be divided into five aspects:

1. Exploration and evaluation of deposits
2. Recovery
3. Extraction of different elements
4. Marketing and economics of the operation
5. Legal problems

Exploration and Evaluation of Deposits.   In the past the occurrence of nodules was determined by photography of the ocean bottom, or by obtaining samples via dredging or coring. Modern methods of exploring for nodules include using deep-sea television cameras and free-falling sampling devices to ascertain the extent and concentration of a particular deposit. The data from these techniques, combined with the many chemical analyses, have shown at least one very promising area in the North Pacific (see Table 5-7). This area is considered large and promising enough for exploitation using present-day technology. Exploration probably will find other promising areas, especially where certain elements are concentrated. Compositional variability could actually be an advantage because, with changing market conditions, the mining locations could be changed to the more profitable nodules.

Recovery.   At present, there appear to be two main methods of recovering manganese nodules from the deep sea: a continuous line bucket system and a hydraulic system using either air or water. The continuous line bucket system was developed by the Japanese and is a purely mechanical system. It consists of a long line or cable that has numerous buckets attached to it (Fig. 5-22). The line is continuous, reaching from the surface ship to the bottom and back to the ship, and as buckets are lowered on one side they are brought up on the other side. A considerable length of the cable is dragged across the sea floor, it is hoped filling the attached buckets with nodules. While this is being done the surface ship is moving perpendicularly and slightly forward to the cable and buckets on the sea floor. In this manner a new portion of the bottom is exposed to the dredging action of the buckets. In 1970 this system was used successfully in a water depth of over 1000 m with 240 buckets each having a capacity of about 50 kg of nodules. Later, apparently successful tests were made in over 4000 m of water. Nevertheless, it seems that hydraulic systems have the most promise for actual use.

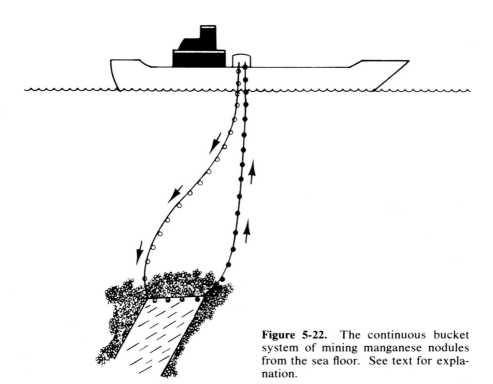

Figure 5-22. The continuous bucket system of mining manganese nodules from the sea floor. See text for explanation.

Hydraulic mining systems, of which there are numerous variations, essentially consist of either a conventional centrifugal pump or a compressed air system and a long pipe reaching to the bottom (Fig. 5-21). Air or water is injected into a dredge on the sea floor, causing a flow of water, sediment, and nodules up the pipe to the surface vessel. Nodules larger than a certain size have to be screened out to prevent clogging of the pipe. In some systems the nodules are crushed before being brought to the surface. Variations of this system are being considered by most of the consortiums previously mentioned.

An imaginative system using a free-falling mining ship has been described by Stechler and Nicholas (1972). In their concept, a special ship would fall free to the bottom. Once on the bottom it would travel along the ocean floor at 2.5 mph (about 4 km/h), filling up with manganese nodules, and then return by floating up to the surface. Each vessel would cost $500 000 and could make eight round trips per day recovering 30 tons per trip. Although most details are proprietary, it sounds overly optimistic.

One problem with the hydraulic and, to a lesser degree, the cable systems method of mining nodules is that of environmental damage. Most economical nodules come from 3000- to 6000-m depth, and the water that comes to the surface with the nodules is generally colder and of a different salinity than the surface waters. In addition, it will contain relatively large amounts of suspended material and probably a higher nutrient content than the surface waters. The National Oceanographic and Atmospheric Administration (NOAA) made a study (Deep Ocean Mining Environmental Study, or DOMES) of mining of manganese nodules, and the potential environmental effects are discussed in Chapter 6, pages 221 – 222. In general, it appears that although mining of nodules has its environmental risks these are considerably fewer than similar mining operations on land.

Extraction of Different Elements.   Preliminary work suggests that physical and smelting methods of separation are not the best and that chemical techniques hold the most promise. In the chemical method the nodules are either partly or completely dissolved and individual elements are removed. The method is especially useful for removing copper and nickel, which are easily taken into solution. Different techniques and reagents can be used to obtain other desired elements.

The specifics of most techniques of refining are kept confidential. Generally, the operation first involves crushing the nodules, followed by drying and leaching with either hydrogen sulfide or hydrogen chloride. In the latter case the important elements become water-soluble chlorides. Individual elements can then be separated by ion-exchange processes and collected on electrolyte cells. A flow diagram of such a process as developed by Deepsea Ventures is shown in Fig. 5-23. Estimated recovery could be as high as 90% of the desired elements. Cost estimates of such a process are about $8–15 per ton of nodules. The cost of setting up such a plant would be immense,

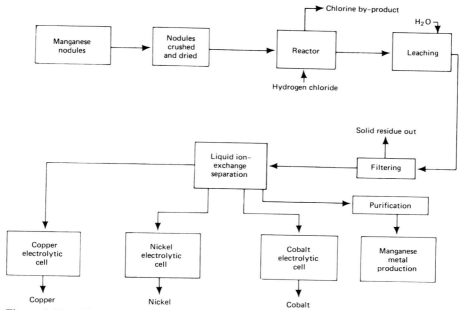

Figure 5-23. Flow diagram of a processing system devised by Deepsea Ventures to mine manganese nodules.

however, being on the order of $50–100 per annual ton of nodule capacity, or $100 million for a plant that handled 1 million tons per year, although, as previously stated, most details are confidential.

Marketing and Economics of the Operation.   There have been numerous estimates of the cost of recovering and refining manganese nodules. These costs, based on numerous assumptions, can be divided into two categories: production costs and capital costs. The estimates for the two generally total between $100 million and $300 million and vary according to methods of mining, methods of refining, amortization schedules, interest charges, and the like. It is hard to argue here, using incomplete data, whether the mining of manganese nodules would produce a profit to those companies engaged in the operation, but it is difficult to imagine them engaging in such a venture if the possibilities of profit were not high.

Perhaps more important is what the effects of mining will have on the world market and the economics of many developing countries. One of the major economic problems associated with mining manganese nodules is that the ratio of metals in the nodules is not in balance with the consumption of these metals by the world's market. This point was made by La Que (1971) for the 1967 world production of these metals and also holds true for present conditions (Table 5-8). It can be seen from Table 5-8 that if nodules were mined at a rate to produce 100% of the world's copper consumption, a con-

Table 5-8. Possible Production of the Different Metals Contained in Maganese Nodules Assuming 100% of the World Production for Each of the Main Metals

Metal	World Production[a] (metric tons)[b]	Pounds of Element Contained in 1 Ton of Nodules[c]	Percentage of Recent World Production of Elements in Nodules That Could Be Obtained from Nodules Assuming 100% World Production of Each of One Element			
			Copper	Nickel	Manganese	Cobalt
Copper	5 822 456	22.5	100	7.2	16	1.5
Nickel	475 988	25.6	1391	100	224	22
Manganese	20 573 518	493	620	44	100	9.8
Cobalt	22 606	5.5	6296	452	1015	100

[a]Data from U.S. Geological Survey (Albers 1973) and latest data available.
[b]2205 pounds = 1 metric ton.
[c]Using data from siliceous ooze area in North Pacific (see Table 5-7).

siderable excess of cobalt, manganese, and nickel would result. If mining were aimed at just 10% of the cobalt market, only a modest amount of nodules could be mined (about 900 000 tons), with a limited amount of other elements produced. It is possible that the increased availability of certain elements from nodules could produce new uses for the metals, but probably not enough to balance the inequalities in the ratios in the nodule. It should be noted that the United States, the major force in the mining of manganese nodules from the deep sea, imports about 15% of its copper, 71% of its nickel, 98% of its manganese, and 98% of its cobalt for an annual trade deficit of about $1 billion. These elements are of strategic importance to the United States and are not stockpiled to any considerable degree.

Another important aspect is that many developing countries depend upon exports of the elements found in manganese nodules for a major source of their foreign revenue. Large-scale mining of manganese nodules therefore could have a considerably detrimental effect on the economics of these countries. In addition, the increased availability of these elements from an ocean mining operation should also reduce the price per pound for these elements. One possible solution could be only to use nodules to meet projected increased demands of the various metals. This solution would not be completely satisfactory because meeting an anticipated 6% increase in the use of copper would result in producing over 300% of the world's demand for cobalt.

Legal Problems.   The main legal problem about the mining of manganese nodules concerns the ownershop of these deep-sea resources. This point was indirectly covered in the 1958 Geneva Conference on the Law of the Sea. Article 1 from the Convention on the Continental Shelf states that the continental shelf is "the seabed and subsoil of the submarine areas adjacent to the coast but outside of the area of the territorial sea, to a depth of 200 m, or beyond that limit, to where the depth of superadjacent waters admits the exploitation of the said areas and to the seabed and subsoil of similar submarine areas adjacent to the coasts of islands." This point of exploitability would theoretically apply to manganese nodule deposits in the deep sea. However, the recent desire of many countries to expand their territorial claims and the suggestion that the resources of the ocean are the "common heritage of all mankind" makes a deep-sea mining venture legally confusing. The present series of Law of the Sea Conferences (see Chapt. 3) is trying to establish international and legal guidelines for the resources of the high seas. The articles concerning sea-bed mining have changed somewhat with each meeting of the Law of the Sea delegates. One article that arose at the 1977 meeting would require mining companies to transfer their privately developed technology to an international sea-bed authority in exchange for mining rights. The desire among many of the developing countries is for the nodules to be mined completely under the control of the authority (who would set price and production controls) and for profits to be equally distributed. The United States has favored a parallel type of system whereby the authority and private industry would operate together. Industries (mainly United States companies in consortiums with other countries) are hesitant to risk further funds on an actual marine mining operation because there is no guarantee that regulations accepted in Law of the Sea negotiations would not result in financial or other obligations to their operation. The problem is especially relevant to the United States because it has the technology to exploit these resources. There is a strong sentiment in the United States to proceed unilaterally with a mining operation with Federal guarantees against interference in international waters or from economic liabilities resulting from the Law of the Sea negotiations.

Because of the uncertainty about the ownership and mining of manganese nodules in the deep sea, the mining industry and its supporters have sought aid in the United States Congress. Several bills have been presented but at the time of this writing none has been passed. If one were to win approval, it could provide assurances to the mining companies for their mining site and give some financial guarantees toward possible losses. Implementation of such laws would also put strong pressure on the Law of the Sea negotiations, perhaps permanently dividing the conference. Alternatively, the United States' position, allowing free participation by private industry, might be better received. This one issue is probably the most intensely argued one at the Law of the Sea Conference.

The suggestion that any deep-sea mining ventures be taxed or licensed and such income be divided among the world's countries as part of their common heritage runs into some problems when actual numbers are used. The total value of those major metals present in manganese nodules (i.e., the value of the manganese, cobalt, nickel, and copper produced) is on the order of $10 billion for 1980. Making the optimistic assumption that 10% of these metals could come from nodules and that a 10% tax was applied, about $100 million would be available for distribution; or about $715 000 for each of the member countries of the Unted Nations. In this light the advantages of sharing the resources from the deep sea do not appear very impressive.

In summary, the mining of manganese nodules from the deep-sea floor will be technologically feasible in the 1980s and sufficient commercial interest and refining techniques presently exist to almost insure that a mining operation will occur, although serious questions concerning legal aspects still exist. No legal regime presently exists to adequately define and protect a deep-sea mining operation, but the Law of the Sea Conference should make some progress on this point. Regardless of the legal confusion concerning the deep sea, some companies might still proceed. An important question about the mining of the nodules concerns which elements are extracted and how they are likely to affect some of the developing countries who are dependent on the exports of these elements for a large source of their foreign currency. Likewise, will private industry be able to operate freely in the deep sea, or will all mining be under the control of a deep-sea authority dominated by countries with no or little mining or technology experience?

Biologic Resources of the Ocean

The biologic resources of the ocean are of considerable value to the human race, both now when many suffer from poor diet and in the future as our population grows. It is thought by many that the oceans will help solve many of the world's food problems but, as we shall see, this possibility is small unless there are major long-range changes in how we use the ocean. At present, the ocean really only plays a small part in supplying the world's food, certainly much less than that of the land.

The basic food problem in the world is that about 2 billion people, half the world's population, have a protein-deficient diet. The definiency is worse in underdeveloped or less developed countries, and often their populations are growing faster than their ability to produce adequate food supplies. The main need is for animal protein, a commodity present in large quantities in the ocean. Fish products, unfortunately, account only for a small portion of the world's total human intake of this protein and much of this amount comes indirectly by our consumption of poultry and livestock that has been fed fish meal products. Even if fish harvest from the ocean were doubled it would not meet the global deficiency of animal protein, but it would help. Some have suggested that increased amounts of protein could be obtained by harvesting plankton, which is the largest numerical group of organisms (except for bacteria) in the ocean, although individual organisms are usually microscopic in size. To obtain large quantities of plankton an immense quantity of water must be filtered (as much as 1 million pounds of sea water to obtain 1 lb of plankton). There are better reasons, however, such as taste, that plankton are not a suitable source of protein. To find a usable source of protein, one must look to organisms higher in the food chain, such as fish, and to new techniques of harvesting them.

Biologic resources of the ocean, like those of the land, are directly related to the production of organic matter by plants. Phytoplankton (floating, generally microscopic plants) are the principal plant in the ocean and are at the base of the food chain. These plants must stay in the upper layers of the ocean because of their need for light for photosynthesis. Nutrients are the other major ingredient needed for photosynthesis; the supply of nutrients is controlled, in large part, by water circulation. Nutrients can be especially abundant in areas of coastal upwelling, where deeper nutrient-rich waters are brought to the surface (Fig. 2-14).

The worldwide production of organic material by phytoplankton (Fig.

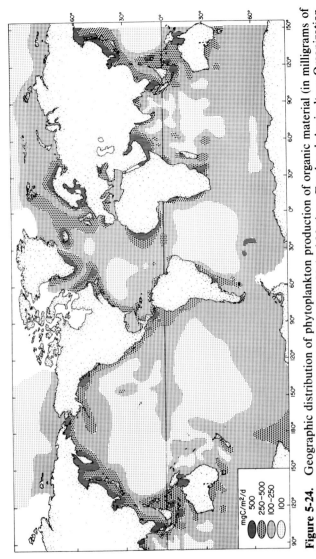

Figure 5-24.   Geographic distribution of phytoplankton production of organic material (in milligrams of carbon per square meter per day).   (Adapted from United Nations Food and Agriculture Organization 1972)

5-24) shows four distinct zones within the ocean:

1. Upwelling areas, such as those off Peru, California, and western Africa, where organic production is extremely high
2. The shallow waters of most continental shelves, which generally have high production
3. The Antarctic, Arctic, and equatorial waters, which are moderately productive because the water is mixed by currents and winds
4. The deep-ocean areas far from land which are essentially biologic deserts

The distribution of zooplankton, which feed on the phytoplankton, is similar. Likewise, locations of the world's major fishing areas follow the pattern of the phytoplankton production of organic material, especially for the near-shore parts of the ocean (Fig. 5-25). This pattern is reasonable because the fish, either directly or indirectly, by eating zooplankton, feed on the organic material produced by phytoplankton. In general, the deep sea contains a relatively small amount of fish or other biologic resources.

An important study concerning organic production in the ocean was published by John Ryther in 1969. He divided the ocean into three main provinces (Table 5-9) based on production of organic material (he lumped

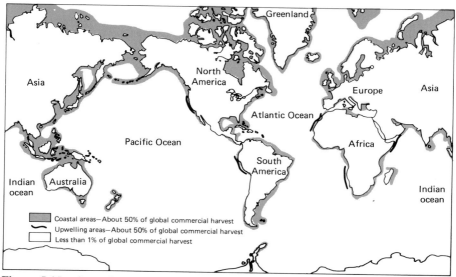

Figure 5-25. Location of the major fishing areas of the world. Shaded areas are coastal areas (about 50% of global commercial harvest); dark "bars" are upwelling areas (about 50% of global commercial harvest); unshaded open-ocean areas produce less than 1% of the global commerical harvest. (From *Patterns and Perspectives in Environmental Science:* Report Prepared for the National Science Board, National Science Foundation, 1972)

Table 5-9. Production of Organic Material in the Ocean and Estimated Amounts of Fish Production[a]

Province	Percent of ocean	Mean Productivity of Organic Material (g carbon/m²/year)	Total Primary Production (billion tons organic carbon)	Feeding Levels[b]	Efficiency of Conversion (%)	Total Fish Production (tons fresh weight)
Open ocean	90	50	16.3	5	10	1 600 000
Coastal zone (including offshore areas of high production)	9.9	100	3.6	3	15	120 000 000
Upwelling zone	0.1	300	0.1	1.5	20	120 000 000

[a] Adapted from Ryther (1969).
[b] Feeding levels refer to number of stages from phyloplankton to harvestable fish product.

the previously mentioned zones 3 and 4 together). It can be noted from the study that although the open ocean makes up about 90% of the ocean, it produces less than 1% of the fish caught and has little potential for increased production. The coastal zone and upwelling provinces occupy 9.9% and 0.1%, respectively, of the ocean but each produce about 50% of the world's fish catch. One of the main reasons for this is that in the open ocean there are more different feeding levels necessary to go from microscopic phytoplankton to an animal large enough to be used by humans than there are in the other two provinces. In going from one feeding level to another, as much as 90% of the organic material may be lost. In addition, the efficiency of conversion (rate of growth of the animal relative to the amount of food consumed) is greater in the coastal zone and upwelling provinces than in the open ocean.

Fisheries

World Fishery Catch

The total amount of fish, crustaceans, mollusks, and other aquatic plants and animals that can be harvested from the sea is subject to considerable debate. The world catch of fish for 1976 was 161.6 billion pounds or about 73.4 million metric tons (Table 5-10). Most of this catch came from off Asia and was mainly used for food (Table 5-11). The most common catch was the small herring, anchovies, and sardines (Table 5-12). Japan caught 14% of the total landings, followed by the USSR (13.8%), China (9.4%), Peru (5.9%), Norway (4.7%) and the United States (4.1%).

There has been considerable debate among marine biologists and fishery experts over what is the maximum sustainable yield from the sea, i.e., how much can be removed before the breeding stock is dangerously reduced. Some feel that we are presently very close to maximum yield, whereas some, perhaps overoptimistically, feel that we have a long way to go and that yields several times the present catch are possible. Any additional increase in yield from the ocean, if possible, will require more sophisticated methods of fishing. It will be necessary for the fishermen to change from "hunters," which they essentially are now, to "herders." In other worlds, it will be necessary for them to be able to control or influence the movement of fish. Aquaculture is one such method of control. Modern electronic devices (Fig. 5-26) permit the modern fisherman to find schools of fish. The fish, however, are generally still caught by conventional techniques, using nets from small ships (Figs. 5-27 and 5-28).

United States Fishery Catch

As an example of the fishing industry within a developed country, I present some data concerning the United States. In 1977 the United States commercial fishery and shellfishing landings were a record $1.5 billion for a

Table 5-10. World Commercial Catch of Fish, Crustaceans, Mollusks, and Other Aquatic Plants and Animals (Except Whales and Seals), by Countries, 1972–1976[a]

Country	Live Weight (thousand metric tons)				
	1972	1973	1974	1975[b]	1976
Japan	10 272	10 748	10 805	10 524	10 620
USSR	7 757	8 619	9 236	9 936	10 134
Peoples Republic of China (Peking)	6 880[c]	6 880[c]	6 880[c]	6 880[c]	6 880[c]
Peru	4 725	2 328	4 145	3 447	4 343
Norway	3 186	2 987	2 645	2 550	3 435
United States	2 695[d]	2 719[d]	2 744[d]	2 743[d]	3 004[d]
Republic of Korea	1 341	1 684	2 023	2 133	2 407
India	1 637	1 958	2 255	2 328	2 400
Denmark	1 443	1 465	1 835	1 767	1 912
Thailand	1 679	1 679	1 516	1 553	1 640
Spain	1 536	1 578	1 510	1 523	1 483[c]
Indonesia	1 270	1 265	1 336	1 382	1 448
Philippines	1 128	1 251	1 298	1 366	1 430
Chile	818	691	1 158	929	1 264
Canada	1 169	1 157	1 037	1 029	1 136
South Vietnam	978[c]	1 014[c]	1 014[c]	1 014[c]	1 014[c]
Iceland	726	902	945	995	986
Brazil	602	704	765	836[c]	950[c]
France	797	823	808	806	806
North Korea	800[c]	800[c]	800[c]	800[c]	800[c]
Poland	544	580	679	801	750
Bangladesh	640[c]	640[c]	640[c]	640[c]	640[c]
Republic of South Africa	664	710	648	636	638
Namibia (S.W. Africa)	527[c]	710[c]	840[c]	761[c]	574[c]
Mexico	459	479	442	499	572
England and Wales	539	557	534	497	520
Malaysia	359	445	526	474	517
Scotland	530	562	538	468	514
Burma	453	463	434	485	502
Nigeria	446	466	473	478	495
Federal Republic of Germany	419	478	526	442	454
Italy	430	399	431	417	420
Senegal	294	316	357	363	361
Faeroe Islands	208	246	247	286	342
Portugal	452	482	436	375	339
Netherlands	348	344	326	351	284
Argentina	238	302	296	229	282
All others	7 132	7 377	7 172	7 150	7 171
Total	66 121	66 808	70 300	69 893	73 467

[a]Source: Food and Agriculture Organization of the United Nations (FAO), *Yearbook of Fishery Statistics*, 1976, Vol. 42.

[b]Revised.

[c]Data estimated by FAO.

[d]Includes the weight of clams, oysters, scallops, and other mollusk shells. This weight is not included in other United States catch statistics.

Table 5-11. Estimated Disposition of World Commercial Catch (Except Whales and Seals), 1975 and 1976[a]

Item	Percent of Total	
	1975	1976
Marketed fresh	27.2	26.5
Frozen	17.3	18.1
Cured	12.0	11.3
Canned	13.6	13.3
Reduced to meal and oil	28.5	29.4
Miscellaneous purposes	1.4	1.4
Total	100.0	100.0

[a]Source: Food and Agriculture Organization of the United Nations (FAO), *Yearbook of Fishery Statistics*, 1976, Vol. 43.

Table 5-12. World Commercial Catch of Fish, Crustaceans, Mollusks, and Other Aquatic Plants and Animals (Except Whales and Seals), by Species Groups, 1973–1976[a]

Species Group	Live Weight (thousand metric tons)			
	1973	1974	1975	1976
Herring, sardines, anchovies, *et al.*	11 314	13 888	13 618	15 089
Cods, hakes, haddocks, *et al.*	11 970	12 699	11 882	12 116
Freshwater fishes	9 293	9 244	9 599	9 532
Miscellaneous marine and diadromous fishes	8 676	8 382	8 021	8 445
Jacks, mullets, sauries, *et al.*	5 740	5 454	5 935	7 389
Redfish, basses, congers, *et al.*	4 320	4 865	5 071	4 950
Mollusks	3 459	3 424	3 779	3 917
Mackerels, snoeks, cutlass-fishes, *et al.*	3 418	3 611	3 590	3 340
Tunas, bonitos, billfishes, *et al.*	1 999	2 125	1 976	2 209
Crustaceans	1 932	2 009	1 979	2 080
Miscellaneous aquatic plants and animals	1 461	1 559	1 336	1 396
Flounders, halibuts, soles, *et al.*	1 248	1 175	1 143	1 123
Shads, milkfishes, *et al.*	759	743	755	697
Salmon, trouts, smelts, *et al.*	554	500	552	555
Sharks, rays, chimaeras, *et al.*	589	544	571	533
River eels	52	55	58	64
Sturgeons, paddlefishes, *et al.*	24	25	27	31
Total[b]	66 808	70 300	69 893	73 467

[a]Source: Food and Agriculture Organization of the United Nations (FAO), *Yearbook of Fishery Statistics*, 1976, Vol. 42.
[b]Figures may not add to total because of rounding.

DEPTH IN FEET OR FATHOMS

Thermal gradients

Concentrations of fish

Bottom

a

b

Figure 5-26, a and b. Echo sounding record (**a**) that is used by fishermen (**b**) to locate concentrations of fish. A moderately large school of fish is seen near the bottom of the record. (Photograph courtesy of Raytheon Marine Company)

a

Figure 5-27, a and b. a Small purse seining fishing boat. The seine is a long net with floats on top and weights on the bottom. It is laid around a school of fish and a purse line on the bottom of the net is pulled tight, like a purse, trapping the fish (**b**) inside it. (Photographs courtesy of John Mason, Woods Hole Oceanographic Institution)

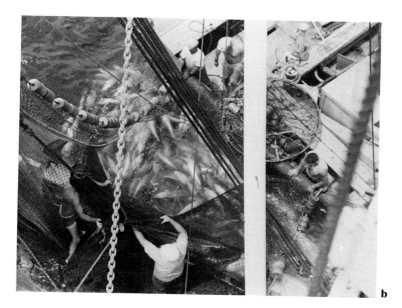

b

catch of 5.2 billion pounds. The catch was about 200 000 lb less than the previous year (United States fishery statistics are published yearly by the National Oceanographic and Atmospheric Administration, or NOAA). Edible species made up 2.9 billion pounds, worth $1.4 billion. So-called nonedible species, used for making fish meal and other industrial purposes, totaled 2.3 billion pounds and were worth about $111 million.

In 1977 the United States imported a record total of $2.6 billion of edible and industrial fish, an increase of 12% over the previous year's imports. The edible fish products made up the major amount of the total used by the United States, being 2.2 billion pounds. Exports were $520 million, which was

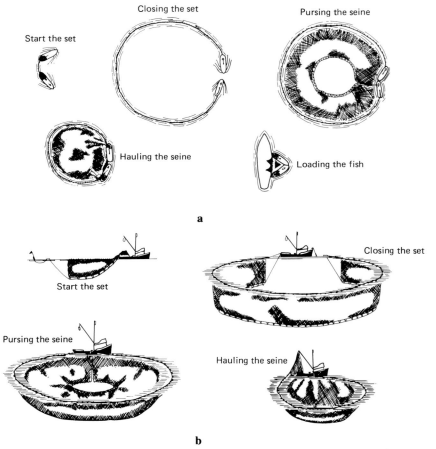

Figure 5-28, a—e. Some methods of catching fish. **a** Two boats and mothership method of purse seining; **b** one-boat purse seine operation; **c** otter trawl gear; **d** pair trawl; **e** dredging operation. (Adapted from various sources)

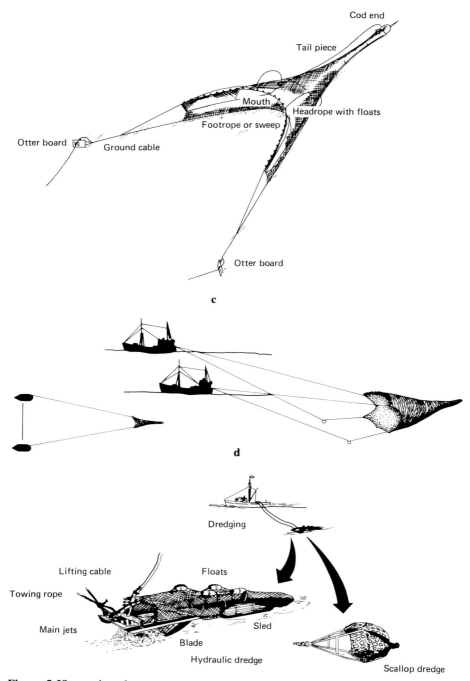

Figure 5-28, continued.

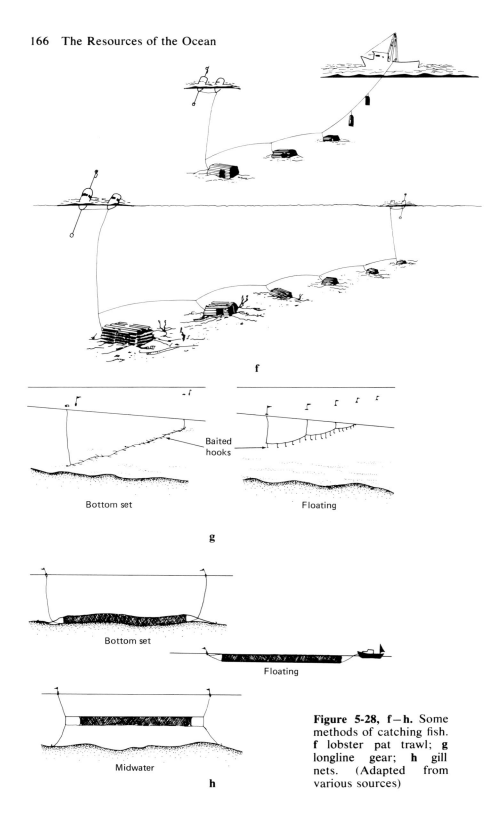

Figure 5-28, f–h. Some methods of catching fish. **f** lobster pat trawl; **g** longline gear; **h** gill nets. (Adapted from various sources)

an increase of 35% over 1976. The total United States supply of commercial fish products in 1977 (landings plus imports) was 10.6 billion pounds (Fig. 5-29), a 10% decrease over 1976. Of this, total edible fish and shellfish accounted for 7.4 billion pounds. There is a very important point shown in Fig. 5-29 which is that the United States fishery catch has hardly changed in the last decade—some of the possible reasons for this are discussed below.

Of the United States commercial catch, about 60% was taken in United States territorial waters (0–3 naut. mi.). The total foreign and domestic catch off the United States (out to 200 naut. mi.) was 8.5 billion pounds, of which the United States share rose to 56% from 48% in 1976. It should be noted that this was the first year that foreign fisheries (out to 200 naut. mi.) were controlled under the Fisheries Conservation and Management Act of 1976 (discussed below). The amount of fish caught by United States recreational fishermen is hard to determine. The most recent data, from 1970, estimate a salt-water catch of about 1.6 billion pounds or about 35% of the total catch at that time by commercial fishermen. A survey of boats made in 1973 found that there were 8 million privately owned recreational fishing boats and that slightly over 1 million were used in salt water.

United States per capita consumption of fish products in 1977 was 12.8 lb per person (does not include catch by recreational fishermen), slightly less than the record 13.0 lb consumed in 1976. For a comparison, in 1977 the

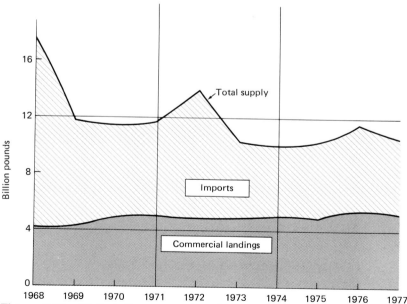

Figure 5-29. United States supply of edible and industrial fishery products, 1968–1977 (billion pounds, round weight). (From United States Department of Commerce 1977)

consumption of beef, pork, lamb, and veal in the United States was 154.8 lb per person, and poultry consumption was 52 lb per person. This relatively small consumption rate of fish products is similar to that of other developed countries (see Table 5-13, which is in live-weight equivalents).

Table 5-13.   Annual Per Capita Consumption of Fish and Shellfish, by Region and Country, 1972–1974 Average[a]

Region and country[b]	Estimated Live-Weight Equivalent	
	Kilograms	Pounds
North America		
United States	15.7	34.6
Canada	16.3	35.9
Latin America		
Argentina	6.1	13.4
Barbados	28.9	63.7
Bolivia	1.4	3.1
Brazil	6.7	14.8
Chile	16.9	37.3
Columbia	4.1	9.0
Costa Rica	4.2	9.3
Cuba	19.5	43.0
Dominican Republic	5.2	11.5
Ecuador	7.2	15.9
El Salvador	3.1	6.8
Guatemala	1.0	2.2
Guyana	18.5	40.8
Haiti	1.4	3.1
Honduras	1.1	2.4
Jamaica	29.5	65.0
Mexico	4.6	10.1
Nicaragua	3.7	8.2
Panama	11.3	24.9
Paraguay	1.2	2.6
Peru	15.6	34.4
Puerto Rico	25.4	56.0
Surinam	22.6	49.8
Trinidad and Tobago	10.1	22.3
Uruguay	5.3	11.7
Venezuela	11.5	25.4
Europe		
Albania	1.7	3.7
Austria	7.6	16.8
Belgium and Luxembourg	18.3	40.3
Bulgaria	7.7	17.0

Table 5-13.   Annual Per Capita Consumption of Fish and Shellfish, by Region and Country, 1972–1974 Average[a] (*continued*)

Region and country[b]	Estimated Live-Weight Equivalent	
	Kilograms	Pounds
Europe (*continued*)		
Czechoslovakia	7.7	17.0
Denmark	34.7	76.5
Finland	22.1	48.7
France	21.5	47.4
German Democratic Republic	19.4	42.8
Republic of Germany	11.5	25.4
Greece	15.2	33.5
Hungary	4.5	9.9
Iceland	66.0	145.5
Ireland	11.2	24.7
Italy	12.6	27.8
Malta	12.4	27.3
Netherlands	13.2	29.1
Norway	47.3	103.6
Poland	16.3	35.9
Portugal	58.5	129.0
Romania	6.3	13.9
Spain	38.5	84.9
Sweden	31.0	68.3
Switzerland	10.7	23.6
United Kingdom	18.9	41.7
Yugoslavia	3.2	7.1
USSR	26.6	58.6
Near East		
Afghanistan	.1	.2
Cyprus	6.1	13.4
Egypt	3.2	7.1
Iran	.5	1.1
Iraq	2.5	5.5
Israel	15.3	33.1
Jordan	2.2	4.9
Lebanon	3.3	7.3
Libya	4.9	40.8
Saudi Arabia	6.4	14.1
Sudan	1.3	2.9
Syria	1.3	2.9
Turkey	4.7	10.4
Yemen Arab Republic	1.4	3.1
Yemen (Aden)	23.6	52.0

Table 5-13.   Annual Per Capita Consumption of Fish and
Shellfish, by Region and Country, 1972–1974 Average[a]
(*continued*)

Region and country[b]	Estimated Live-Weight Equivalent	
	Kilograms	Pounds
Far East		
Bangladesh	11.6	25.6
Burma	13.0	28.7
Sri Lanka (Ceylon)	8.7	19.2
China (Peking)	8.3	18.3
Hong Kong	50.6	111.6
India	2.9	6.4
Indonesia	9.6	21.2
Japan	68.8	151.7
Cambodia (Kampuchea, Democratic)	11.2	24.7
North Korea	28.6	63.1
Republic of Korea	34.2	75.4
Laos	6.3	13.9
Malaysia		
Sabah	41.3	91.1
Sarawak	31.3	69.0
West Malaysia	22.5	49.6
Nepal	.2	.4
Pakistan	1.1	2.4
Philippines	30.3	66.8
Singapore	48.1	106.0
Thailand	22.0	48.5
Vietnam	25.2	55.6
Africa:		
Algeria	2.0	4.4
Angola	7.4	16.3
Benin	12.7	28.0
Burundi	3.1	6.8
Cameroon	13.6	30.0
Central African Empire	6.4	14.1
Chad	15.8	34.8
Congo (Brazzaville)	23.4	51.6
Ethiopia	.6	1.3
Gabon	13.6	30.0
Gambia	24.5	54.0
Ghana	28.1	61.9
Guinea	2.9	6.4
Ivory Coast	25.4	56.0
Kenya	2.6	5.7
Liberia	16.6	36.6

Table 5-13. Annual Per Capita Consumption of Fish and Shellfish, by Region and Country, 1972–1974 Average[a] (*continued*)

Region and country[b]	Estimated Live-Weight Equivalent	
	Kilograms	Pounds
Africa (*continued*)		
Madagascar	6.2	13.7
Malawi	8.3	18.3
Mali	9.6	.2
Mauritania	21.9	.3
Mauritius	16.1	.5
Morocco	4.8	10.6
Mozambique	3.1	6.8
Niger	2.1	4.6
Nigeria	6.8	15.0
Rhodesia	2.9	6.4
Senegal	33.8	74.5
Sierra Leone	24.6	54.2
Somalia	.3	.7
Republic of South Africa	9.4	20.7
Tanzania	11.4	25.1
Togo	11.3	24.9
Tunisia	5.8	12.8
Uganda	15.7	34.6
Upper Volta	1.1	2.4
Zaire	9.0	19.8
Zambia	13.1	28.9
Oceania		
Australia	14.1	31.1
New Zealand	15.9	35.1

[a]Source: Food and Agriculture Organization of the United Nations (FAO).
[b]Note: Data for most countries are tentative.

There are many reasons why the United States fishing industry has not improved its share of the total catch over the last few years. One factor was that foreign fishermen used to take about twice the amount of fish that American fishermen caught from the waters adjacent to the United States coast. In 1976, however, the U.S. Fishery Conservation and Management Act was passed which restricted foreign fishing in a 200-mile zone off the United States (effective 1 March 1977). This act, discussed below, reduced foreign fishing and appears to have made more fish available to United States fishermen. Another factor for the low catch by United States fishermen is the lack of financial incentive to the fishing industry. In many other countries food shortages have encouraged governmental support for the fishing industry. The United States, for various reasons, does not seem to mind importing the fish it needs. Sometimes the imported fish are cheaper than the locally caught ones; this can arise because of the better fishing techniques, lower labor costs, etc., of foreign fishermen. A lower price for foreign fish in turn depresses the price that the United States fishermen receive, perhaps preventing them from improving their techniques or equipment.

There are other factors, such as laws and customs, that have effectively reduced the catch of American fishermen. A good example of this is fish protein concentrate (FPC), a tasteless, odorless, cheap flour made from processed fish (including species not usually eaten; also, almost all parts of the fish are used). Although the product is an excellent source of protein and without any taste indication of its fish origin, it has never been able to overcome the generally conservative custom of people toward new foods and become acceptable in the United States diet. Clearly, social and cultural barriers and customs must be considered in developing fish products. Two other examples are eels and squid, both of which are delicacies in Europe but are hard even to give away in the United States. An example of a law that often reduces the United States catch is that any fishing boat that unloads fish in the United States must have been built in the United States. As a result, many United States fishing boats are inferior and often cost more than foreign-built craft.

Fishing Regulations

Fishing regulations, although usually imposed with good intentions, have often been very cumbersome and ineffective. For example, fishing for haddock off the northeastern United States used to be somewhat regulated by the International Commission for Northwest Atlantic Fisheries (ICNAF). Under their ineffective jurisdiction, haddock catch dropped from over 130 million pounds in 1966 to less than 5 million in 1972.

Fish management is a very difficult operation, because there are strong local, state, national, and foreign pressures whereby each group would like to maximize their catch and often minimize their effort. Fish, of course, do not respect territorial boundaries, and little is known of the important biologic parameters concerning fish, such as what is their preferred environment, what produces fastest growth, what promotes successful breeding, and how similar species interact and compete. Until recent years no mechanisms existed to enforce or establish conservation principles on an international level; the exceptions were several international agrreements but, as mentioned above, they often were not very effective. The same situation exists on a state or local level. Local fishermen are especially vulnerable to foreign "invasions" into their waters, whereas distant-water fishermen (such as tuna fishermen off South America) are more tolerant, because they often are the "invaders." The fishing industry itself, because of its nature, makes management extremely difficult—fish can be taken by any fisherman and

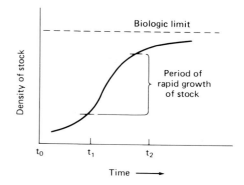

Figure 5-30. Growth of a fish stock. (Adapted from Knight, 1973)

there is often no motivation for individual fishermen to restrict the size of their catches. The feeling is that one should take as much as one can (i.e., fish are common property belonging to anybody), because if one does not someone else will. This has led to intense fishing, and probably overfishing of certain preferred species, such as flounder, which in some instances may severely reduce the viable reproductive population of the species.

Fishing economics in some aspects are not very complicated. The fisherman wants as much money (i.e., fish) as possible but there are limiting biologic factors. A typical growth pattern for a fish stock (Fig. 5-30) initially begins with the number being low as the population starts to get established; as more mature fish become available for reproduction, there is a period of rapid growth. The growth and increase in numbers cannot continue indefinitely because of lack of food, lack of space, predation, etc., and eventually the rate of increase of the population levels off. This total number is variable because other factors, such as natural mortality and environmental conditions, influence the size of the population. In addition, it is difficult to obtain all the data to construct an accurate assessment.

There is in a fishery a point where a maximum sustainable yield (MSY) is possible (see Fig. 5-31); past this point the value or catch per fishing effort decreases. An analysis of catch versus fishing effort (Fig. 5-31) shows that

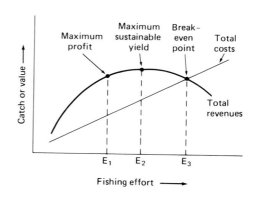

Figure 5-31. Some aspects of fishing catch or fishing value as compared to fishing effort (i.e., number of boats, days at sea, etc.). (From Knight, 1973)

maximum profits (E_1) are generally produced at less than MSY for the fishery. Fishing beyond the MSY will cause overfishing for the resource and eventually will be unprofitable (past E_3). With adequate regulations fishing can be controlled so that point E_3 is not reached.

In the United States there was no national fisheries plan until the passage of the Fishery Conservation and Management Act in 1976. Prior to this, the United States had belonged to as many as 15 international fishing agreements. The aim of most of these agreements was to regulate and control fishing within certain areas or for certain species. Generally there was a feeling that the waters of the United States were often being overfished by foreign countries using large and sophisticated vessels. This view, although simplistic, was often correct and led to a protective view toward United States offshore fisheries. Similar positions have been considered by some countries during the Law of the Sea Conference. It should be emphasized that most of the world's fishing catch comes from areas that would fall within a 200-mile economic resource zone if one were established on a worldwide basis.

The United States Fishery Conservation and Management Act

The Fishery Conservation and Management Act (P.L. 94-265) provides for a national fisheries program and the conservation and management of fishery resources within a defined fishery conservation zone (FCZ). The FCZ extends from the seaward end of the territorial sea to 200 naut. mi. from shore. For most states the seaward end of the territorial sea starts 3 naut. mi. from shore; for Texas and the Gulf Coast of Florida it starts at 3 leagues, or 9 naut. mi. Within the territorial sea individual states have control over fishing. In the new area of extended jurisdiction the United States accepted exclusive management for all fish, except for certain highly migratory tunas. The act also provides for exclusive management authority over the fishery resources of the continental shelf and authority over anadromous species beyond the FCZ (except for when they are within another nation's territorial sea or fishery conservation zone, if it is recognized by the United States). The act also established clear rules controlling access by foreign countries into the FCZ as well as national guidelines for United States fishermen.

If a foreign country wishes to fish within the United States 200-mile zone it must negotiate an agreement with the U.S. Department of State (with the cooperation of NOAA). The agreement is called a Governing International Fishery Agreement (GIFA). For a foreign country to negotiate a GIFA it must agree to the conditions of the Fishery Conservation and Management Act; if so, the request is then transmitted by the President to Congress for review. After a GIFA is entered into, each foreign country must submit (to the U.S. Department of State) an application for each vessel, describing the vessel and giving an anticipated catch by species and the area in which it plans to work. This information is then transmitted to Congress, the Coast

Guard, the appropriate Regional Fishery Council (discussed in a later section), and the Director of the National Marine Fishery Service (NOAA). The input and recommendations of the various groups are sent to the Director of the National Marine Fishery Service (NMFS) and he or she, after discussions with the State Department and Coast Guard can approve the application. If approved, the information is transmitted by the State Department to the foreign country. A series of fees is charged, which include a permit fee based on the tonnage of the vessels involved; a poundage fee, which is based on the United States dockside price for the appropriate fish; and an observer fee, whereby the foreign nation pays for all costs of putting observers on board the foreign vessel. Eighteen countries entered into GIFAs during the first year of the act. In 1977 the United States collected approximately $11 million in fees for foreign fishing within the FCZ.

For the United States fisherman, the act established eight Regional Fishery Management Councils, which are to prepare plans for fishery management within their regions (these plans must be approved by the Secretary of Commerce) for both commercial and recreational fishing. The councils may also recommend regulations and comment on fishing permits requested by foreign countries within their region. The councils are rotated so that each year one-third of the members can be replaced. The members are selected by the Secretary of Commerce from a list of nominees submitted by the governors of the applicable states. Many members of the councils are state fisheries directors and people with financial interests in fishing and processing. There have been considerable differences in the styles of operation of various councils.

Any fishery management plan developed is required to be consistent with a series of seven national standards that were established in the act. These are as follows (Dept of Commerce, 1977):

1. Conservation and management measures shall prevent overfishing while achieving, on a continuing basis, the optimum yield from each fishery.

2. Conservation and management measures shall be based upon the best scientific information available.

3. To the extent practicable, an individual stock of fish shall be managed as a unit throughout its range, and interrelated stocks of fish shall be managed as a unit or in close coordination.

4. Conservation and management measures shall not discriminate between residents of different States. If it becomes necessary to allocate or assign fishing privileges among various United States fishermen, such allocation shall be (a) fair and equitable to all such fishermen; (b) reasonably calculated to promote conservation; and (c) carried out in such manner that no particular individual, corporation, or other entity acquires an excessive share of such privileges.

5. Conservation and management measures shall, where practicable, promote efficiency in the utilization of fishery resources; except that no such measure shall have economic allocation as its sole purpose.

6. Conservation and management measures shall take into account and allow for variations among, and contingencies in, fisheries, fishery resources, and catches.

7. Conservation and management measures shall, where practicable, minimize costs and avoid unnecessary duplication.

It is premature to completely evaluate the effects of the act, although some good and bad has resulted, but the net results appear beneficial. The implementation of the act has clearly made the United States take a definite position concerning its marine biologic resources and how best to use them. It has developed methods whereby these resources are shared and managed. Input from conservationists, the public, and consumers (both foreign and domestic) is given a chance to be considered in policy decisions. The act will also lead to the collection of improved data, biologic, economic, and social, so that better management schemes can be prepared. The effects on the United States domestic industry should be considerable, and for the first time in a decade the United States fish catch should start to improve. For example, haddock and cod catch increased 110% and 42%, respectively, off New England in the first year of the act. This, however, has not been without its problems, as many United States fishermen were dismayed to find that there were times when fishing for certain species was restricted.

The unilateral action of the United States in declaring a 200-mile fishery zone was not lost on the delegates to the Law of the Sea Conference and this action has led to other countries' introducing similar legislation; however, it is possible that they might have done so even without the action by the United States. The act does have provisions for amendments according to the Law of the Sea results. It is easy to argue that if all countries were to implement fair and equitable management programs, the mutual benefit for all would increase considerably.

Other Factors Influencing Fish Catch

Fishing success is also influenced by factors beyond the control of fishermen or politicians. Changing oceanographic or atmospheric variables can dramatically influence fishing success. One of the best examples of this occurred off Peru, where for several years the largest fishery of the world had existed. This is an area of intense upwelling mainly caused by prevailing southerly winds. Occasionally, however, atmospheric conditions change, causing more northerly winds (winds from the north); this causes the warming of the surface waters and a slackening of the upwelling. The result is that the available nutrients are quickly used up, organic production by phytoplankton decreases, and associated marine and bird life dependent on it dissappear or die. This changing wind condition, called "El Niño" (or The Child, because it periodically occurs around Christmas time), has had a disastrous effect on Peruvian fisheries. In 1971, Peru had an annual anchovy catch of over 10 million metric tons, which made it the world leader in fishing. An El Niño condition occurred in 1972 and Peru's catch dropped to 4.5 million tons, and even further in 1973 to 1.8 million tons. Part of this decline also resulted from overfishing in previous years. The result was disastrous to the Peruvian economy as well as to other countries of the world who were dependent on the anchovies for fish meal or other products. The diaster also affected the guano industry (fertilizer from the droppings of marine birds that also feed on the anchovies). In recent years the fishery has recovered somewhat, but it has never reached its former size.

In may countries there are social customs and religious barriers that affect fish consumption. Among Catholics in the United States and elsewhere, it was a custom to eat fish on Friday. Some people therefore ate fish only on Friday and, as this custom was modified, the consumption of fish dropped considerably. In other places there just is not a general interest in eating fish (Table 5-13). Likewise, in some countries problems of storage and transportation combined with social aspects strongly limit fish consumption.

A major problem facing the fishing industry is that of pollution. Many fish and other biologic resources, such as clams, mussels, and some crabs, breed in estuaries that are areas especially vulnerable to pollution. In addition, increased levels of mercury, DDT, and other pollutants have been found in some marine organisms (this is discussed in more detail in Chapt. 6).

Opportunities to Increase Food Production from the Ocean

In recent years the rate of increase of the world's fish catch has slowed down (Table 5-10), and as previously stated, some feel we may be getting near our maximum sustainable yield from the ocean. There are other opinions that it is possible to increase our harvest from the ocean. It seems clear that although fishing catches may be somewhat improved by increasing fishing efforts (more ships fishing more days), any major increase must come from an innovative method or by tapping a previously unused or underused resource. It should be emphasized that many areas of the ocean, especially those parts far from land, are essentially biologic deserts and, without some modification of their environment, are never likely to become major areas of fish production. There are four methods that could be used to increase the amount of fish products.

The first of these is by better fishing techniques, including improved management, or simply better handling and transportation to keep spoilage to a minimum. Fishermen are essentially hunters, but if through the use of electronic techniques they could attract fish, catch per unit effort would be improved considerably. Japanese scientists have experimented with using sound to lure fish into nets or other collecting devices, with some success. Likewise, it may be possible to train one variety of fish to help find or catch other species. The generally close association of porpoises with tuna is one obvious possibility. Another possibility is using chemicals to attract fish to a particular area. Researchers have had some success in luring salmon to certain streams by chemicals (see Chapt. 9, page 303).

It is questionable whether all the major fishing areas of the world are actually being fished. Some regions, such as off parts of Africa and in the southern hemisphere, could probably safely yield higher catches of fish if the proper equipment were available and if the local population were interested in obtaining the fish. This, likewise, is true for many of the world's rivers, lakes, and coastal areas.

Electronic techniques to locate fish have been commonly used on foreign ships for years and are becoming popular in the United States (Fig. 5-26). More sensitive devices would improve fishing success. Perhaps most important among this category of ways to improve fish catch is that of fishery management. The proceeding pages have emphasized how proper management could reduce pressure on certain species and keep catch to an optimum.

A second technique that could aid in increasing the harvest from the ocean is to obtain and use so-called scrap or underutilized species. In some cases one person's trash fish is another's delicacy. For example, shark and squid are popular in the Far East but rarely eaten in the United States. Another example would be increased use of FPC (fish protein concentrate).

In addition to trash fish, it should also be possible to harvest organisms lower in the food chain that are more abundant than the predators at the top of the food chain. There have been experiments in harvesting plankton, such as diatoms, but the costs and its generally unappealing aspect suggest that it may not make a viable food product. Probably a better possibility is that of using the large shrimp—such as forms called krill—that are extremely abundant in the Antarctic, where they form the major food for some species of whales (Fig. 5-32). Krill have a high protein content and appear to be extremely abundant. One estimate is that their total mass is over 10 trillion pounds (higher than the mass of the human race). It has also been suggested that up to 200 million tons of krill could be harvested yearly without hurting the population (others argue for a much smaller harvest). Such a catch is about triple that of all other forms in the ocean. Krill (*Euphausia superba*) are 2–3 in. (5–7 cm) long and stay in very large schools or clumps that makes catching them fairly simple. The West Germans have developed a technique that allows krill to be harvested at rates of 8–12 tons per hour. The large amount of krill could be related to the recent decline of the whale population in the Antarctic area, who feed on them.

The use of krill does have problems, one of which is that they decompose very rapidly and must be processed almost immediately after they are caught. The Japanese have developed a technique whereby the krill is im-

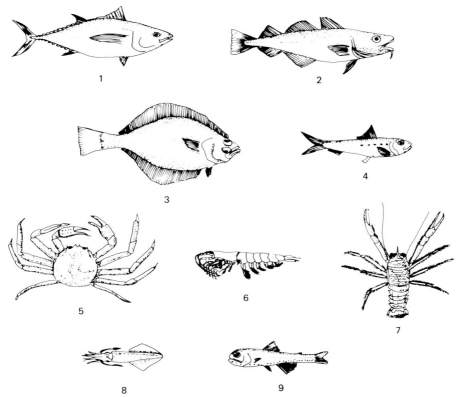

Figure 5-32. Some conventional and unconventional fish species: 1, tuna; 2, cod; 3, flounder; 4, sardine; 5, crab; 6, euphausiid; 7, red crab; 8, squid; and 9, lantern fish. (Adapted from Alverson 1975)

mediately processed aboard ship into a beanlike paste that can be marketed or mixed with batter or cheese. Its flavor is similar to shrimp.

The concentration of krill in Antarctic waters, far from most countries, combined with the need of processing as well as harvesting facilities, should make fishing in this area difficult and relatively expensive. However, if the size of the resource is sufficient, it may be well worth the expense. There could be some difficulties with "ownership" of the area, because the Antarctic is currently an international area where no claims are presently recognized, based on the Antarctic Treaty of 1961. The treaty put a 30-year moratorium on land claims and military and commercial development in the area. Within recent years there has been interest among some of the 12 nations signatory to the treaty and others in developing a new treaty to permit the catch and management of this potential resource. Interestingly (and perhaps by plan), Antarctica has not been among the present items of discussion at the Law of the Sea Conference.

There are other species, although not as dramatic in numbers as krill, that could be harvested. One is the lantern fish (Fig. 5-32), which is widespread throughout the ocean and often occurs in large schools at midwater depths. Along with other pelagic species, this relatively small fish could be harvested at a yield of perhaps 100 million metric tons per year, according to some estimates. It should be emphasized that little is known about the biology of these fish or of krill, so estimates may be overly optimistic. Other underutilized species are red crab and varieties of seaweed.

Seaweed is an important product from the sea that should have even more uses in the future. At present about 1.7 million tons are harvested throughout the world and it has a market value of close to half a billion dollars. Most of the seaweed is eaten, especially in the Far East where it may constitute as much as 25% of some countries' diets. Seaweed contains several trace elements as well as important vitamins and minerals.

Seaweed is also used as fertilizer and as an animal feed. Some forms of seaweed are used for the polysaccharides, agar, algin, and carrageenan they contain. These compounds have important stabilizing, thickening, gelling, and emulsifying characteristics and are very commonly utilized in foods and chemicals. They are used in toothpaste, candy, ice cream, soap, and paint to name a few products. Agar is also used for the culture of bacteria and has many other medical uses. Seaweed products and their potential uses easily exceed supply and seaweed cultivation is a new and growing industry. Seaweed culture can often be easily done in coastal regions or be part of a larger aquaculture system.

The third method to increase food production from the ocean is by increasing or stimulating the primary production of organic matter by plants via the addition of nutrients. Several innovative techniques have been suggested, such as bringing up nutrient-rich deep water as part of ocean thermal energy systems (see Chapt. 9, pages 287-288) and using it to increase the growth of phytoplankton and subsequently of animals, such as shellfish, that feed on the phytoplankton. Likewise, the wise disposal of sewage or other nutrient-rich waste products into marshes or parts of the coastal zone could lead to similar results. Experiments on increasing productivity of coastal marshes by the use of secondarily treated sewage have proceeded for several years, but it has not yet been shown that the resulting food products (shellfish, etc.) can be safely consumed. Another similar aspect is using the heat from power plants to increase the growth of certain species, such as lobsters.

The fourth technique for increasing production is aquaculture. Aquaculture or, as the seawater system is sometimes called, mariculture, can be defined as the growing of freshwater or marine organisms under controlled conditions. This technique, which is commonly practiced in many areas of the world, is in its relative infancy in the United States (Table 5-14). Its importance in such countries as China, India, and Japan is considerable, and

Table 5-14. Estimated World Production of Fish, Shellfish, and Seaweeds by Aquaculture in 1975[a]

	Metric Tons		Metric Tons
Finfish	3 980 492	Burma	1 500
China (all provinces		El Salvador	1 208
excluding Taiwan		Canada	1 103
Province)	2 200 000	Greece	900
Taiwan, Province of	81 236	Chile	800
China		Uganda	700
India	490 000	Singapore	680
USSR	210 000	Kenya	400
Japan	147 291	Nepal	400
Indonesia	139 840	Venezuela	332
The Philippines	124 000	Switzerland	300
Thailand	80 000	Ireland	207
Bangladesh	76 485	Republic of Korea	169
Nigeria	75 000	The Netherlands	129
Poland	38 400	Ecuador	90
Republic of Vietnam	30 000	Central African Empire	43
Yugoslavia	27 000	Cyprus	40
Romania	25 000	Ghana	40
Hungary	23 515	Zambia	29
United States	22 333	Paraguay	23
Italy	20 500	Ivory Coast	10
Madagascar	17 392	Puerto Rico	9
Democratic Republic of		Shrimps and prawns	15 663
Germany	16 000	India	4 000
France	15 000	Indonesia	4 000
Czechoslavakia	12 222	Thailand	3 300
Isreal	12 169	Japan	2 779
Denmark	12 120	Ecuador	900
Brazil	12 000	Taiwan, Province of	
Federal Republic of		China	549
Germany	8 900	Singapore	105
Sri Lanka	7 659	Republic of Korea	30
Egypt	7 000	Oysters	591 386
Mexico	7 000	Japan	229 899
Malaysia	6 559	United States	129 060
Zaire	5 000	France	71 448
Cuba	4 500	Republic of Korea	56 008
Hong Kong	4 019	Mexico	45 000
Norway	3 500	Thailand	23 000
Austria	2 500	Taiwan, Province of	
United Kingdom	2 000	China	13 359
Finland	1 940	Australia	9 200
Belgium	1 800	Canada	5 080
Tanzania	1 500	United Kingdom	3 000

Table 5-14. Estimated World Production of Fish, Shellfish, and Seaweeds by Aquaculture in 1975[a] (*continued*)

	Metric Tons		Metric Tons
Spain	2 289	Taiwan, Province	
The Netherlands	1 500	of China	13 898
Chile	870	The Philippines	33
The Philippines	782	Scallops	62 600
New Zealand	700	Japan	62 600
Senegal	191	Cockles and other mollusks	29 987
Mussels	328 517	Malaysia (cockles)	28 000
Spain	160 000	Taiwan, Province	
The Netherlands	100 000	of China	1 243
Italy	30 000	Republic of Korea	733
France	17 000	The Philippines	11
Federal Republic of		Seaweeds	1,054 793
Germany	14 000	Japan	502 651
Republic of Korea	5 578	China (all provinces	
Chile	1 260	excluding	
Yugoslavia	287	Taiwan Province)	300 000
The Philippines	182	Taiwan, Province of	
New Zealand	150	China	7 347
Tunisia	60	Republic of Korea	244 795
Clams	38 851	Total	6,102 289
Republic of Korea	24 920		

[a]Data from Pillary (1976).

the total aquaculture harvest is equal to almost 10% of the world's fish catch and is more than the United States total. In Israel, for example, as much as 75% of the fish sold comes from pond aquaculture. The most popular fish is carp, followed by mullet. Many of the ponds are used in conjunction with chicken farms, and the waste products from the farms are used as a source of nutrients for plants in the ponds. In Japan oysters are grown that reach marketable size within a year. Some Japanese oyster rafts have produced 4 tons of oyster meat in a year, which is among the highest rates of production of animal protein per unit area yet achieved (Table 5-15).

Aquaculture has been practiced for as many as 4000 years in areas of the Middle and Far East. In the United States close to 30% of the landing of Pacific salmon and over half of Columbia River salmon were released from hatcheries. Likewise, close to half of United States oysters and catfish come from aquaculture facilities; seaweed is also a common product of such facilities.

In oyster farming, the baby oysters, called spat, float and move with the ocean currents until they encounter a shell on the bottom to which they can become attached. Oyster fishermen—actually oyster farmers is a better term—place shells on the bottom in environments, such as bays and inlets, suitable for growth. The spat, after attaching to the shells, will grow until large enough to be harvested by the "farmers." Another method, common

Table 5-15. Comparison of Natural Production in the Ocean with Some Aquaculture Results

Area	Product	Method	Production (lb/acre/year)
Chesapeake Bay[a]	Various fish	Natural	125
North Sea[a]	Various fish	Natural	25
Peru Current[a]	Various fish	Natural	330
Phillipines[b]	Milkfish	Fishpond aquaculture	500
France[b]	Gray Mullet	Fishpond aquaculture	300
British Columbia[b]	Oysters	Raft aquaculture	40 000
British Columbia[b]	Mussels	Raft aquaculture	400 000

[a]Data from McHugh (1967).
[b]Data from Quayle (1972).

in Japan, is to suspend shells on wires in the water (see Fig. 5-33); thus, more spat can grow in a particular area. This method, a sort of three-dimensional farming, has several other advantages: It removes the baby oysters from their natural enemies, the starfish, and places the oysters where more food is available. Bottom-living oysters can feed only on the plankton that

Figure 5-33. Scallop shells (attached to strings) being used for the settlement of oyster spat. (Photograph courtesy of Woods Hole Oceanographic Institution)

settle to the bottom, whereas oysters living on suspended shells can feed on the much larger quantity of plankton that float by in the currents. The technique can also be adopted for mussels.

Lobsters may also do very well in an aquaculture system, especially systems that use heated water. Selected lobsters grown in warm sea water in a Massachusetts lobster hatchery reached sexual maturity in 2 years, whereas it usually took 8 years in their natural environment. This result raises the possibility of using heated water from perhaps a nuclear power plant to increase growth of selected organisms.

It is estimated that worldwide aquaculture production could reach 50 million metric tons, and perhaps 1 million metric tons in the United States. United States aquaculture facilities are relatively small or experimental; catfish, oysters, and crayfish presently are the only profitable edible crops, but shrimp and lobster may soon join this category. Seaweed, baitfish (e.g., minnows), and worms, although not edible, are also profitable aquaculture products. Emphasis in the United States has been on more exotic species high in the food chain, rather than on various forms of fish (exceptions are trout and salmon). A 1978 National Academy of Sciences report on aquaculture (U.S. National Research Council 1978) suggested that because of the "rapidly changing situation in the United States with respect to cost, availability and acceptability of aquatic foods it appears to justify the parallel development of new initiatives . . . aquaculture, involving concepts, approaches, and species not yet utilized on a commercial basis in the United States." They recommended that the new programs include:

1. Cultivation of aquatic plants for food, feed, chemical products, waste treatment and biomass for conversion to energy (see Chapt. 9, pages 000–000)

2. Closed-cycle aquaculture—sort of essentially self-contained food factories

3. Polyculture, using several species that occupy different ecologic niches

4. Ocean ranching

5. Alternatives to current feeding practices

Many of these recommendations will require research into the biologic aspects of the organisms concerned as well as into engineering design. Some of these systems can be developed with ocean thermal energy conservation (OTEC) systems (see Chapt. 9, pages 281–287) or other innovative future uses of the sea.

Several systems using the waste products of a sewage system (Fig. 5-34) are polyculture in nature. In many systems the products that are to be eaten have not yet been shown to be sufficiently clean of the pollutants in the sewage, but other products, such as seaweed, are adequate for use. In the system shown in Fig. 5-34 the phytoplankton remove most of the nutrients in

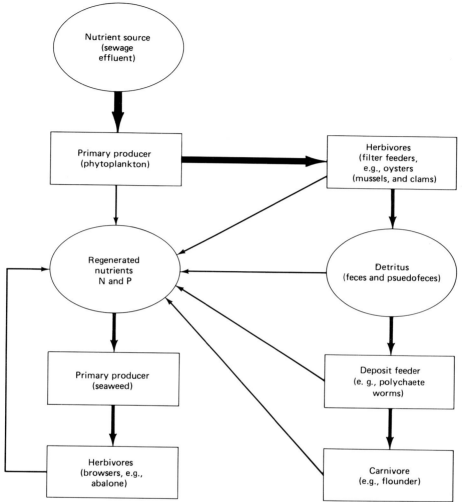

Figure 5-34.   Flow diagram of a tertiary sewage treatment–marine aquaculture system. (Ryther *et al.* 1972b)

the water and are then fed to herbivores, such as oysters; waste products are eaten by deposit feeders, such as worms, or are put back into the system and removed via seaweed growth or by other herbivores. The net result besides a biologic product is a relative cleaning of sewage water. Problems in the system include the fate of viruses and the accumulation of heavy metals or other organic pollutants by some of the organisms.

Generally, when aquaculture is practiced in the marine environment it is done in nearshore areas, such as bays and estuaries. However, it may, be possible to adapt the system for open-ocean use.   Open-ocean aquaculture will allow access to larger and more varied sources of food and generally solve the problem of accumulation of waste products, which can be an important detriment in a restricted system.   An open-ocean system has problems of predator control and possible loss or escape of some of the organisms.   It also will present some challenging engineering problems.   One technique is to use already present offshore platforms, such as drilling rigs or production platforms, to grow oysters or mussels.   More imaginative ideas could be self-contained floating systems or ones as part of offshore energy systems.

In conclusion, it seems that by using some or all of the above techniques that it might be possible to double or even triple the catch or harvest from the ocean.   Within future years, therefore, a total harvest approaching 200×10^6 metric tons is conceivable.   This value could be equivalent to about 50 lb of fish product per person (assuming a doubling of the population).   Assuming that 10% is annual protein there is still not enough to supply the world's need (8–16 lb/year/person of animal protein are needed).   Nevertheless, it could be a considerable help.   A major flaw in this calculation is that many of the fish products are not dirctly eaten but used as fish meal for animal feed (at a considerable loss of efficiency in protein transfer).   In addition, the problem still exists of getting the protein to the people who need it most.   This is often an economic, cultural, or social problem.

References

Adams, M.V., C.B. John, R.F. Kelly, A.E. LaPointe, and R.W. Meurer, 1975, *Mineral Resource Management of the Outer Continental Shelf.* U.S. Geol. Survey Circular 720, 32 pp.

*Albers, J.P., 1973, Summary Petroleum and Selected Mineral Statistics for 100 Countries, including Offshore Areas.   United States Geological Survey Professional Paper 817.

Alexander, L., and V. Norton, 1977, The 200-mile limit III: Maritime problems between U.S. and Canada: *Oceanus,* vol. 20, p. 24–34.

*Alverson, D.L., 1975, Opportunities to increase food production from the world's oceans: *Marine Technology Society Jour.,* vol. 9, p. 33–40.

Alverson, D.L., 1977, The 200-Mile Limit II: The North Pacific Fishery Management Council, *Oceanus,* Vol. 20, p. 18–23.

Alverson, D.L., A.R. Longhurst, and J.A. Gulland, 1970, How much food from the sea? *Science,* vol. 168, p. 503–505.

*American Geological Institute, 1953, *Glossary of Geology and Related Sciences,* Washington, D.C., American Geological Institute.

Amos, A.F., C. Garside, K.C. Haines, and O.A. Roels, 1972, Effects of surface-discharged deepsea mining effluent: *In* Horn, D.R., Ed. *Ferromanganese De-*

posits on the Ocean Floor. Harriman, New York: Arden House and Lamont-Doherty Geological Observatory, p. 271–282.

Antarctic Problems: Tiny krill to usher in new resource era, 1977, *Science*, v. 196, p. 503–505.

Arrhenius, G., 1974, *Science*, v. 183, p. 502–503.

Arrhenius, G., 1975, Mineral Resources on the Ocean Floor: *Technol. Rev*, March/April, p. 22–26.

Auer, P.L., 1974, An integrated National energy research and development program: *Science*, v. 184, p. 295–301.

*Bäcker, H., and M. Schoell, 1972, New deeps with brines and metalliferous sediments in the Red Sea: *Nature Phys. Sci.*, v. 240, p. 153–158.

Bardach, J.E., 1968, Aquaculture: *Science*, v. 161, p. 1098–1106.

Bardach, J.E., 1969, Aquaculture: Amplification and correction: *Science*, v. 163, p. 493–494.

Bardach, J.E., J. Ryther, and W.O. McLarney, 1972, *Aquaculture*. New York: John Wiley and Sons, 868 pp.

Berg, R.R., J.C. Calhoun Jr., and R.L. Whiting, 1974, Prognosis for expanded U.S. production of crude oil: *Science*, v. 184, p. 331–336.

Berryhill, H.L. Jr., 1974, *The Worldwide Search for Petroleum Offshore—A Status Report for the Quarter Century, 1947–72*, U.S. Geol. Survey Circular 694, 27 pp.

*Bischoff, J.L., and F.T. Manheim, 1969, Economic potential of the Red Sea heavy metal deposits: *In* Degens, E.T., Ross, D.A., Eds., *Hot Brines and Recent Heavy Metal Deposits in the Red Sea*. New York: Springer-Verlag, p. 535–541.

Bisko, D., J.R. Dunn, and W.A. Wallace, 1969, *Planning for Non-renewable Resources in Urban-Suburban Environs*. Report for the U.S. Bureau of Mines, Troy, N.Y.: Rennselaer Polytechnic Institute, p. 1–16.

Bonatti, E., T. Kraemer, and H. Rydell, 1972, Classification and genesis of submarine iron–manganese Deposits: *In* Horn, D.R., Ed., *Ferromanganese Deposits on the Ocean Floor*. Harriman, N.Y.: Arden House and Lamont-Doherty Geological Observatory, p. 149–166.

Booda, L.L., 1976a, Fishery management and enforcement face formidable complications: *Sea Technol.*, June, p. 12–15.

Booda, L.L., 1976b, Marine mining faces bright economic future: *Sea Technol.* August, p. 23–28.

*Bostrom, K., and M.N.A. Peterson, 1966, Precipitates from hydrothermal exhalations on the East Pacific Rise: *Econ. Geol.*, v. 61, p. 1258–1265.

Branco, R., 1973, Rational development of seabed resources: Issues and conflicts: *Ocean Management*, v. 1, p. 41–54.

Breaux, J., 1977, What is the future for deepsea mining? *Sea Technol.*, July, p. 41.

Brett, J.R., J.R. Calaprice, R.V. Ghelardi, W.A. Kennedy, D.B. Quayle, and C.T. Shoop, 1972, *A Brief on Mariculture*. Fisheries Research Board of Canada: Pacific Biological Station, Nanaimo, B.C., 46 pp.

Browder, J.A., 1977, *The Fishery Act of 1976: A Summary; The Management Councils: A Description*. Univ. of Miami Sea Grant Special Report No. 12, 16 pp.

*Cronan, D.S., 1967, Geochemistry of Some Manganese Nodules and Associated Pelagic Deposits. Ph.D. Thesis, Imperial College, University of London.

*Cronan, D.S., 1972, Composition of Atlantic manganese nodules: *Nature Phys. Sci.,* v. 235, p. 171–172.

*Cronan, D.S., and J.S. Tooms, 1969, The geochemistry of manganese nodules and associated pelagic deposits from the Pacific and Indian Oceans: *Deep-Sea Research,* v. 16, p. 335–359.

Davenport, J.M., 1971, Incentives for ocean mining: *Marine Technology Scoiety Jour.,* v. 5, p. 35–40.

*Degens, E.T., and D.A. Ross, 1969, *Hot Brines and Recent Heavy Metal Deposits in the Red Sea.* New York: Springer-Verlag, 600 pp.

*Department of Commerce, 1977, *Fisheries of the United States, 1977,* United States National Oceanographic Administration, National Marine Fisheries Service.

Deuel, D.G., 1977, Marine recreational fisheries—Uses and values: *In Coastal Recreation Resources in an Urbanizing Environment.* A monograph of the Cooperative Extension Service, Univ. Massachussets, U.S. Dept. of Agriculture, and Massachussetts Institute of Technology, p. 34–37.

Dunstan, W.M., and K.R. Tenore, 1972, Intensive outdoor culture of marine phytoplankton enriched with treated sewage effluent: *Aquaculture,* v. 1, p. 181–192.

Edwards, R., and R. Hennemuth, 1975, Maximum yield: Assessment and attainment: *Oceanus,* v. 18, p. 3–9.

*Emery, K.O., 1969, Continental rises and oil potential: *Oil Gas Jour.,* v. 67, p. 231–243.

Emery, K.O., 1974a, *Report on World Resources of Fossil Fuels.* Committee for the Mineral Resources and Environment, Washington D.C., National Academy of Science. - Panel III, 32 pp.

*Emery, K.O., 1974b, Provinces of promise: *Oceanus,* v. 27, p. 14–19.

Emery, K.O., 1975, New opportunities for offshore petroleum exploration: *Technol. Rev.,* March/April, v. 77 p. 31–33.

Emery, K.O., 1976, Offshore oil: Technology and emotion: *Technol. Rev.,* February, v. 78 p. 31–37.

Emery, K.O., 1977, Mineral deposits of the deep-ocean floor: *Marine Mining,* v. 1, p. 1–71.

Emery, K.O., and C.O'D. Iselin, 1967, Human food from ocean and land: *Science,* v. 157, p. 1279–1281.

*Emery, K.O., E. Uchupi, J.D. Phillips, C.O. Bowin, E.T. Bunce, and S.T. Knott, 1970, Continental rise off eastern North America: *Amer. Assoc. Petrol. Geol. Bull.,* v. 54, p. 44–108.

Extended Fishery Jurisdiction: Problems and Progress, 1977, 1978, Proceedings of the North Carolina Governor's Conference on Fishery Management under Extended Jurisdiction, 11–12 October, 1977, 246 pp.

*Food and Agricultural Organization, 1972, *Atlas of the Living Resources of the Sea,* New York, United Nations.

*Food and Agricultural Organization, 1976, *Yearbook of Fishery Statistics, 1976,* vol. 42, New York, United Nations.

Freeman, B.L., 1977, A description of recreational finfishing along the Atlantic coast in relation to the utilization of living marine resources: *In: Coastal Recreation Resources in an Urbanizing Environment,* A Monograph of the Cooperative Extension Service, Univ. Massachussetts, U.S. Dept. of Agriculture, and Massachussetts Institute of Technology, p. 30–33.

Friedman, A.G., 1977a, Deepsea mining legislation: *Marine Policy*, v. 1, p. 341–342.

Friedman, A.G., 1977b, Seabed minerals, the United States and cartelization: A rejoinder: *Marine Technology Scoiety Jour.*, v. 11, p. 38–40.

Government Seeking Ways to Encourage Aquaculture: 1978, *Science*, v. 200, p. 33–35.

Gulland, J.A., 1971, *The Fish Resources of the Ocean*. England: Fishing News (Books), Ltd.,

*Hedberg, H.D., J.D. Moody and R.M. Hedberg, 1979, Petroleum prospects of the deep offshore, *American Assoc. of Petroleum Geologists Bulletin*, vol. 63, n. 3, pp. 286–300.

*Heezen, B.C., and C.D. Hollister, 1971, *The Face of the Deep*. New York: Oxford Univ. Press, 659 pp.

Hess, H.D., 1973, Environmental impact study of offshore sand and gravel mining: *MTS Marine Technology Society Jour.*, v. 7, p. 49–52.

Holt, S.J., 1969, The food resources of the ocean: *Sci. Amer.*, v. 221, p 178–194

Horn, D.R., B.M. Horn, and M.N. Delach, 1972a, Distribution of ferromanganese deposits in the world ocean: *In:* Horn, D.R., Ed., *Ferromanganese Deposits on the Ocean Floor*. N.Y.: Harriman, Arden House, and Lamont-Doherty Geological Observatory, p. 9–18.

Horn, D.R., B.M. Horn, and M.N. Delach, 1972b, Ferromanganese Deposits of the North Pacific. National Science Foundation, Tech. Rep. No. 1, NSF GX-33616-IDOE.

*Horn, D.R., B.M. Horn, and M.N. Delach, 1973, Factors Which Control the Distribution of Ferromanganese Nodules and Proposal Research Vessel's Track North Pacific. National Science Foundation, Tech. Rep. No. 8, NSF GX-33616-IDOE.

Hughes, J.T., J.J. Sullivan and R. Shleser, 1972, Enhancement of lobster growth: *Science*, v. 177, p. 1110–1111.

Hurley, P.M., 1975, Plate tectonics and mineral deposits: *Technol. Rev.*, March/April, v. 77 p. 15–21.

*Knight, G.H., 1973, International Law of Fisheries, Louisiana State University Teaching Aid, Issue No. 2 - Center for Wetlands Resources.

Lampe, H.C., N. Marshall, J.G. Sutinen, L.O. Vidaeus, and D.T. Westin, 1974, *Prospects for Fisheries Development Assistance*. Marine Tech. Rept. Ser. No. 19, International Center for Marine Resource Development, University of Rhode Island, Kingston R.I., 41 pp.

Landsberg, H.H., 1974, Low-cost, abundant energy: Paradise lost? *Science*, v. 184, p. 247–253.

Lane, A.L., 1977, U.S. Needs in marine minerals: *Marine Tech. Soc. Jour.*, v. 11, p. 30–36.

*La Que, F.L., 1971, Prospects for and from deep ocean mining: *Marine Technol. Soc. Jour.* v. 5, p. 5–15.

Library of Congress Congressional Research Service, 1976, *Effects of Offshore Oil and Natural Gas Development on the Coastal Zone*. A Study Prepared for the Ad Hoc Select Committee on Outer Continental Shelf, House of Representatives, 396 pp.

Liston, J., and L. Smith, 1977, Fishing and the fishing industry: *In: International Transfer of Marine Technology—A Three-Volume Study*. J. Kildow, principal investigator. Massachusetts Institute of Technology Sea Grant Rept. 77-20, p. 143–202.

Lyman, H., 1977, The 200-mile limit I: The New England Regional Fishery Management Council: *Oceanus*, v. 20, p. 7–17.

Manheim, F.T., 1972a, Composition and origin of manganese–iron nodules and pavements on the Blake Plateau: *In:* Horn, D.R., Ed., *Ferromanganese Deposits on the Ocean Floor;* Harriman, N.Y.: Arden House and Lamont-Doherty Geological Observatory, p. 191–204.

Manheim, F.T., 1972b, *Mineral Resources Off the Northeastern Coast of the United States.* U.S. Geol. Survey Circular 669, Washington, D.C.

Mann, K.H., 1973, Seaweeds: Their productivity and strategy for Growth: *Science*, v. 182, p. 975–981.

McCaslin, J.C., 1974a, Offshore oil production soars: *Oil Gas Jour.*, v. 72, p. 136–142.

McCaslin, J.C., 1974b, Tomorrow's oil will be found offshore: *Oil Gas Jour.*, v. 72, p. 95.

McHugh, J.L., 1967, Estuarine nekton: *In:* Lauff, G.H., Ed, *Estuaries.* Amer. Assoc. Advan. Sci. Publ. No. 83, Washington, D.C., p. 581–620.

*McKelvey, V.E., and F.H. Wong, 1970, *World Subsea Mineral Resources.* U.S. Geol. Survey Misc. Geol. Invest. Map I-632.

McKelvey, V.E., F.H. Wong, S.P. Schweinfurth, and W.C. Overstreet, 1969, Potential mineral resources of the United States outer continental shelves: *In: Public Land Law Review Commission Study of Outer Continental Shelf Lands of the United States,* Vol. 4. (Appendices.)

*Menard, H.W., and S.M. Smith, 1966, Hypsometry of ocean basin provinces: *Jour. Geophys. Res.*, v. 71, p. 4305–4325.

*Menard, H.W., and J.Z. Frazer, 1978, Manganese nodules on the sea floor: Inverse correlation between grade and abundance: *Science*, v. 199, p. 969–971.

*Mero, J.L., 1965, *The Mineral Resources of the Sea.* New York: Elsevier.

*Mero, J.L., 1972, Potential economic value of ocean floor manganese nodule deposits: *In:* Horn, D.R. Ed. *Ferromanganese Deposits on the Ocean Floor.* Harriman, N. Y.: Arden House and Lamont-Doherty Geological Observatory, p. 105–106.

*Miller, A.R., C.D. Densmore, E.T., Degens, J.C. Hathaway, F.T. Manheim, P.F. McFarlin, R. Pocklington, and A. Jokela, 1966, Hot Brines and Recent Iron Deposits in Deeps of the Red Sea: *Geochim. Cosmochim. Acta*, v. 30, p. 341.

Milliman, J.D., O.H. Pilkey, and D.A. Ross, 1972, Sediments of the continental margin of the eastern United States: *Geol. Soc. Amer. Bull.*, v. 83, p. 1315–1334.

Moody, J.D., 1970, Petroleum demands of future decades: *Amer. Assoc. Petrol. Geol. Bull.*, v. 54, p. 2239–2245.

National Academy of Sciences, 1975, *Mining in the Outer Continental Shelf and in the Deep Ocean.* Washington, D.C., 19 pp.

*National Petroleum Council, 1972, *U.S. Energy Outlook,* Washington, D.C., National Petroleum Council.

Office of Technology Assessment, 1977, *Establishing a 200-Mile Fisheries Zone.* Washington, D.C., 132 pp.

Overall, M.P., 1968, Mining phosphorite from the sea: *Ocean Industry,* September, p. 44–52.

*Patterns and Perspectives in Environmental Science, 1972, Report prepared for the National Science Board, National Science Foundation. Washington, D.C.: U.S. Government Printing Office.

Peterson, S.B., 1976, Report of the Workshop on Extended Jurisdiction. Woods Hole Oceanographic Institution Tech. Rept. 76-73, 53 pp.

Pinchot, G.B., 1970, Marine farming: Sci. Amer., v. 223, no 3 p. 15-21.

Quayle, D.B., 1972, Commercial culture of invertebrates, Part I, Molluscan mariculture in British Columbia. In: Brief on Aquaculture. Fisheries Board of Canda, p. 24-32.

Rathjen, W.F., 1975, Unconventional harvest: Oceanus, v. 18, p. 36-37.

Raymond, R.C., 1976, Seabed minerals and the U.S. economy: A second look: Marine Technology Society, v. 10, p. 12-18.

Roels, O.A., J.S. Babb, G.L. Hamm, and K.C. Haines, 1973, Mariculture in an Artificial Upwelling System. Paper presented at Fifth Annual Offshore Technology Conference, Houston, Texas, 29 April-2 May 1973, p. 391-398.

Rona, P.A., 1973, New evidence for seabed resources from global tectonics: Ocean Management, v. 1, p. 145-159.

*Ross, D.A., 1977, Introduction to Oceanography, 2nd Edition, Englewood Cliffs, N.J.: Prentice-Hall.

*Ross, D.A., E.E. Hays and F.C. Allstrom, 1969, Bathymetry and Continuous Seismic Profiles of the Hot Brine Area of the Red Sea: in Degens, E.T. and Ross, D.A., Eds., Hot Brines and Recent Heavy Metal Deposits in the Red Sea, New York: Springer-Verlag, pp. 82-97.

Ross, D.A., R.B. Whitmarsh, S.A. Ali, J.E. Boudreaux, R. Coleman, R.L. Fleisher, R.W. Girdler, F.T. Manheim, A. Matter, C. Nigrini, P. Stoffers, and P. Supko, 1973, Red Sea drillings: Science, v. 179, p. 377-380.

*Ryther, J.H., 1969, Photosynthesis and fish Production in the sea: Science, v. 166, p. 72-76.

Ryther, J.H., 1975, How much protein and for whom? Oceanus, v. 18, p. 10-22.

Ryther, J.H., W.M. Dunstan, K.R. Tenore, and J.E. Huguenin, 1972a, Controlled eutrophication—Increasing food production from the sea by recycling human wastes: Bio-Science, v. 22, p. 144-152.

Ryther, J.H., K.R. Tenore, W.M. Dunstan, J.C. Goldman, J.S. Prince, V. Vreeland, W.B. Kerfoot, N. Corwin, J.E. Huguenin, and J.M. Vaughn, 1972b, The Use of Flowing Biological Systems in Aquaculture, Sewage Treatment, Pollution Assay, and Food-Chain Studies. Progress Report, NSF-RANN GI-32140, 1 January-31 December, 1972.

Schaefer, M.B., and D.L. Alverson, 1968, World fish potentials: In: The Future of the Fishing Industry in the United States. Univ. of Washington Publ. in Fisheries Service, Vol. IV, Seattle, Washington.

*Science, 1974, Manganese Nodules (I): Mineral Resources on the Deep Seabed, Science, vol. 183, pp. 502-503.

Science, 1974, Oil and Gas Resources: Did U.S.G.S. Gush Too High? Science, v. 185, p. 127-129.

Science in Europe: Moratorium set on Antarctic oil at October meeting: 1977, Science, v. 198, p. 709-712.

Shepherd, C.J., 1974, The economics of aquaculture—A review: Oceanogr. Marine Biol. Ann. Rev., v. 13, p. 413-420.

Skinner, B.J., and K.K. Turekian, 1973, *Man and the Ocean*. Englewood Cliffs, N.J.: Prentice-Hall, 149 pp.

Smith, W.J., 1972, International control of deepsea mineral resources: *Naval War College Review*, vol. 25, p. 82–90.

Spinelli, J.L., L. Lehman, and D. Weig, 1974, Composition, processing and utilization of red crab (*Pleuroncodes planipes*) as an aquacultural feed ingredient: *Jour. Fishery Res. Board Canada*, v. 31, p. 1025–1029.

*Stechler, B.G., and J.R. Nicholas, 1972, Recovery of deep ocean nodules: "A new approach": *In:* Horn, D.R., Ed., *Ferromanganese Deposits on the Ocean Floor.* Harriman, N.Y.: Arden House and Lamont-Doherty Geological Observatory, p. 141–148.

*Swallow, J.C., and J. Crease, 1965, Hot salty water at the bottom of the Red Sea: *Nature*, v. 205, p. 165.

Terry, O.W., 1977, *Aquaculture, Mesa New York Bight Atlas Monograph 17*, New York Sea Grant Institute, Albany, New York, 36 pp.

Theobald, P.K., S.P. Schweinfurth, and D.C. Duncan, 1972, *Energy Resources of the United States*. U.S. Geol. Survey Circular 650, 27 pp.

Towards Project Interdependence: Energy in the Coming Decade, 1975, Report prepared for the Joint Committee on Atomic Energy, United States Congress, Washington, D.C.: U.S. Government Printing Office, 249 pp.

United Nations Department of Economic and Social Affairs, 1970, *Mineral Resources of the Sea*. New York, 49 pp.

*United Nations Food and Agriculture Organization, 1972, *Atlas of the Living Resources of the Seas,* Department of Fisheries, Rome.

United States Department of Commerce, National Oceanic and Atmospheric Administration, 1974, *Fisheries of the United States, 1973,* Current Fishery Statistics No. 6400. Washington, D.C.: U.S. Government Printing Office.

United States Department of Commerce, National Oceanic and Atmospheric Administration, 1976, *Fisheries of the United States, 1975,* Current Fishery Statistics No. 6900. Washington, D.C.: U.S. Government Printing Office.

United States Department of Commerce, National Oceanic and Atmospheric Administration, 1976 *Fisheries of the United States, 1976,* Current Fishery Statistics No. 7200. Washington, D.C.: U.S. Government Printing Office.

United States Department of Commerce, 1976c, *A Marine Fisheries Program for the Nation*. Washington, D.C., 74 pp.

United States Department of the Interior, 1976, *Leasing and Management of Energy Resources on the Outer Continental Shelf*. Bureau of Land Management/Geological Survey, 40 pp.

*United States Geological Survey, 1974, U.S. Petroleum and Natural Gas Resources, Washington, Dept. of the Interior, March 1974.

United States National Research Council, 1978, *Aquaculture in the United States: Constraints and Opportunities*. Report by the Committee on Aquaculture, Board of Agriculture and Renewable Resources; Commission on Natural Resources, National Research Council, National Academy of Sciences, 123 pp.

Voss, G.L., 1973, *Cephalopad resources of the world:* FAO Fish Circular No. 149, Rome.

*Weeks, L.G., 1968, The gas, oil and sulfur potentials of the sea: *Ocean Industry,* June, p. 43.

Wenk, E. Jr., 1969, The physical resources of the ocean: *Sci. Amer.,* v. 221, n. 3 Sept. p. 167–175.

Woods Hole Oceanographic Institution, 1976, *Effects on Commercial Fishing of Petroleum Development off the Northeastern United States.* Study conducted within Marine Policy and Ocean Management Program of Woods Hole Oceanographic Institution, 80 pp.

Xavier, A., 1976, The Exploitation of deepsea manganese nodules: Progress and prospects: *Maritime Policy Management,* v. 4, p. 33–40.

Chapter 6

Marine Pollution

Introduction

Pollution is not really a use of the ocean, although it often is a result of misuse. Nevertheless, pollution, its causes, and problems must be considered in a discussion of the opportunities and uses of the sea. This is especially true because the oceans are an important present and future source of energy, food, and mineral resources and the continued exploitation of these resources will put considerable environmental pressures on certain areas, especially the coastal zone. In addition to these uses, the ocean is the main thoroughfare of commerce and the principal area for the disposal of many of our industrial and domestic waste products. All these uses are increasing and will correspondingly increase the probability of future environmental damage by pollution.

The oceans contain a large and varied collection of life forms, most of which are intimately dependent upon each other. These forms are in extremely close contact with the environment and, in most instances, cannot adjust to profound environmental changes. The decrease or elimination of one variety of life can often have important effects on other species. Most marine scientists would probably concur that the varied and immense aspects of marine pollution are the most serious present-day problem of the ocean. The problem is made extremely complex because of the lack of background data. For example, in most instances we do not know to what level the ocean is already polluted, what pollutants were present before the industrial revolution, and what are the levels of the pollution that the ocean can tolerate. Likewise, the considerable size of the marine environment and the slowness of some of its physical processes do not allow easy experimentation; neither can we afford to wait to see what results from our present actions. It has been especially surprising to most marine scientists to see how some pollutants have spread throughout the ocean even to its remotest parts. The finding of high levels of hydrocarbons and DDT in the waters and fauna of the Antarctic was not really anticipated. Alternatively, there have been many grim statements concerning pollution that grossly oversimplify the situation or extremely exaggerate it. Doomsday predictions, such as life dying off in the ocean in a few decades from pollution or other ecologic disasters, raise emotions but do little to develop a clear evaluation of the problem.

Within recent years there has been an increased international awareness of ocean pollution problems. This has led to a number of conferences and even some treaties to try to control pollution, especially by ships at sea. These conferences, meetings, treaties, etc., can have beneficial effects, although it may take years or even decades. Some effects of pollution are reversible and it is possible to decrease them by legislation. For example, Woods Hole Oceanographic Institution scientists have noted a drop in the polychlorinated biphenyls (PCB) content in the Sargasso Sea over the last few years. The decrease, which is about 40-fold, appears to have resulted from a ban on industrial sales of PCB that was initiated in 1970–1971 by the United States and Sweden and joined in 1972–1973 by Japan and the rest of Europe.

The total quantity of pollutants introduced into the ocean yearly is not known. Perhaps even more important, little is known of the composition of individual components of pollutants and at what rate or how they degrade the ocean. Some compounds that are very hazardous and toxic on land may not be important ocean pollutants because of short lives in the ocean. Other compounds can be more of a risk in the marine environment than on land; an example is hexachlorobenzene, which persists in the ocean, and organisms tend to accumulate it. There are many important questions concerning marine pollutants, the answers to which are sometimes essentially unknown; these include:

Method of its introduction into the environment

Rate of its introduction into the environment

Chemical composition of the pollutant

Persistence or residence time of the pollutant in the environment

Toxicity or other effects of the pollutant

Degree of its incorporation into the fauna and flora

Ultimate fate of the pollutant

Perhaps our limited awareness of ocean pollution is understandable, because for many centuries it had been thought that the oceans were essentially unlimited in size and could not be harmed or overfished. This view was probably reasonable considering our knowledge at that time; however, we have since found that all bodies of water can be polluted by human activity. Ironically, one problem is to determine whether certain pollutants in the ocean are the result of human activity or simply occur from natural causes. Some pollutants enter the ocean independently of human activities, from natural weathering and erosion of rock and soil and processes associated with sea-floor spreading. It is conceivable that in some instances the input of a potential pollutant from natural causes can be equal to or exceed human input; but because few background data are available, especially prior to the industrial and urbanization revolution, it sometimes is difficult to evaluate human impact exactly.

Pollutants have several pathways that they may travel to reach the ocean or follow once they are in the ocean. The main pathways include:

From the atmosphere. This includes gases, volatile liquids, propellants (now being reduced), hydrocarbons via combustion, liquids or materials sprayed into the atmosphere, such as pesticides, CO_2, and SO_2.

From land. Disposal of waste products, etc.

From rivers. Sewage or effluent from municipal, industrial, or agricultural usage; any spills or accidental discharges; natural runoff.

Directly into the ocean. Dumping of industrial, municipal, or agricultural waste; dredge spoils, direct input via outfalls; bottom paint and wastes from ships.

Once in the ocean material may be absorbed onto small particles and either sink to form part of the sediment, remain suspended, or be consumed by organisms. Some material, because of tubulence in the upper layers of the ocean, may be returned to the atmosphere or remain on the surface as a slick. Other chemicals are dissolved in sea water and remain there for varied periods of time. Most important, however, is the material that gets incorporated into the biologic food system, which sometimes can be concentrated as it travels up the food chain, eventually reaching potentially harmful levels.

One difficulty in discussing marine pollution is its definition. Exactly what is a pollutant, and can a pollutant in one situation actually be a nonproblem or even beneficial in another situation? A United Nations report (The Sea: Prevention and Control of Marine Pollution, Report of the Secretary General, 1971) defined marine pollution as "the introduction by man, directly or indirectly, of substances or energy into the marine environment (including estuaries) resulting in such deleterious effects as harm to living resources, hazards to human health, hindrance to marine activities, including fishing, impairment of quality for use of sea water and reduction of amenities." Note that this definition does not consider pollution from sources other than human, such as natural oil leaks and volcanic eruptions.

The various types of pollutants can be as varied as human technologic achievements but can be categorized (according to the same United Nations report) as:

1. Disposal of domestic sewage, industrial, and agricultural wastes
2. Deliberate and operational discharge of ship-borne pollutants
3. Interference with the marine environment from the exploration and exploitation of marine minerals
4. Disposal of radioactive waste resulting from the peaceful uses of nuclear energy
5. Military uses of the ocean

It is generally thought that the disposal of domestic sewage and agricultural and industrial wastes is the most serious form of pollution, especially for the near-shore area. However, pollution from oil occasionally gets more attention, perhaps because it is so widespread. In the following sections I discuss the major forms of marine pollutants and some of the ways they affect the marine environment.

Domestic, Industrial, and Agricultural Pollution

An extremely large number of varied pollutants, including human sewage, organic material, and pesticides, enter the ocean from domestic, industrial, and agricultural activities. These pollutants can travel by different pathways (Fig. 6-1), including rivers, land runoff, outfalls, and from the atmosphere. Some of the problems associated with such pollution are:

Disease from infectious organisms. This can generally be prevented by disinfecting techniques but some additives used in disinfecting may have dangerous side effects.

Oversupply of nutrients. As discussed in Chapter 2, certain nutrients are critical for plant growth and, if they are oversupplied, plankton growth can be accelerated, a process called eutrophication. Eventually this can result in the complete removal of the dissolved oxygen in the water by the decay of organic matter. This situation will be lethal to most marine life within the environment.

Oxygen-consuming materials. Most organic material, including human wastes, will oxidize in the marine environment. If there is insufficient water mixing and oxygen supply, the available oxygen could be removed, with results as described above.

Toxic chemicals and minerals. Many industrial and agricultural waste

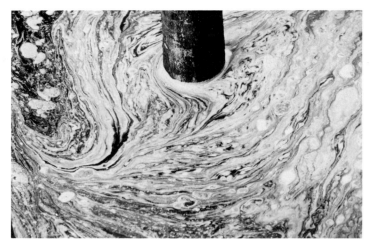

Figure 6-1. Waste water from a pulp mill being discharged into Puget Sound, Washington. (Photograph courtesy of the U.S. Environmental Protection Agency)

products are extremely poisonous or harmful to marine life and in many instances their long-term effects are not known.

Sediments. Rivers carry vast amounts of sediment; for example, United States rivers discharge about 490 million tons of soil and sediment per day into their estuaries and near-shore areas. This material can affect marine life, such as plants, oysters, or clams, as well as sometimes necessitating dredging to keep waterways open. The disposal of the dredged material is an additional pollution problem.

In some of the above, the critical aspect is the effect on the dissolved oxygen in the water. There have been some exaggerations, however, on the possible affects on ocean oxygen content and its implications to humankind. It has been reported that if something decreased or eliminated the oxygen in the ocean that this in turn would drastically affect the atmospheric oxygen content. This conjecture is not correct because the production and consumption of oxygen in the ocean is essentially a self-contained system, with little net exchange with the atmosphere. If all marine photosynthesis were to stop, the atmospheric concentration of oxygen might drop about 10% in 1 million years, a loss that is undesirable but that probably could be tolerated by most life forms.

Synthetic Organic Compounds

The amount of potentially dangerous chemical pollutants has increased considerably in the past few decades because of a large expansion in the chemi-

Table 6-1. Loss of Synthetic Organic Chemicals from Ocean Tankers (Amount of Cargo Residues Discharged at Sea—1970)[a]

Compound	Total Amount Shipped (tons)	Amount Left on Board (tons)
Trichloroethane	4 618	5
Dichloropropane	6 685	Unknown
Perchloroethylene	39 986	36
Ethylene dichloride	241 356	80
Benzene	279 852	241
Toluene	366 441	234
Acrylonitrile	93 453	41

[a]Source: Intergovernmental Maritime Consultative Organization (IMCO), 1973.

cal industry that has produced numerous new products, including synthetic fibers, plastics, and polymers from petrochemical facilities. Likewise, there has been increased use of fertilizers, solvents, and various synthetic and real organic compounds. Many of these compounds and the materials used in their creation will end up in the ocean, either from rivers or from atmospheric input. The fate of these almost countless compounds and their effect on the marine environment, in particular the biologic community, is essentially unknown, but it appears that the ocean may be their ultimate resting ground and they do not degrade very quickly.

Chlorinated and halogenated hydrocarbons are especially dangerous pollutants used in many everyday products; for example, aerosol propellants (recently banned in the United States), flame retardants, fire extinguishers, and many solvents. These compounds can cause liver damage and sometimes may be carcinogenic, among other dangerous effects. The inputs and pathways of many of these compounds to the ocean are varied and generally unknown. Some get directly into the ocean during transportation (Table 6-1), some by industrial accidents, and some even by deliberate dumping.

In some instances marine organisms have shown the potential for biologic concentration or magnification (Fig. 6-2) of some of these substances—i.e., the organisms contain many times the concentration of the compound than is actually present in the water itself.

Among the most important of these synthetic organic compounds may be the polychlorinated biphenyls (PCBs), which are used for electrical insulators, for fire retardants, in making plastics, and in heat exchangers. It has been estimated that of the 1 million tons per year produced, one-fourth, or 250 000 tons, leaks into the environment and a tenth of that, or 25 000 tons, gets into the oceans. Polychlorinated biphenyls have been shown to reduce growth rates of phytoplankton even when present in concentrations as low as 10 parts per billion (ppb). One effect of this could be to change the

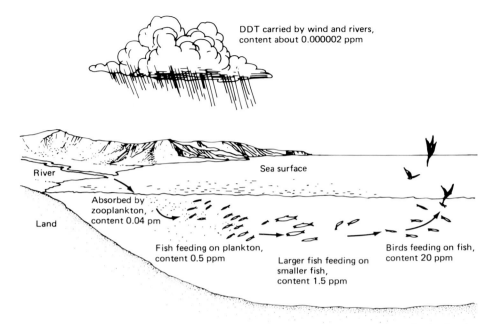

DDT carried by wind and rivers,
content about 0.000002 ppm

Sea surface

River

Land

Absorbed by
zooplankton,
content 0.04 pm

Fish feeding on plankton,
content 0.5 ppm

Larger fish feeding on
smaller fish,
content 1.5 ppm

Birds feeding on fish,
content 20 ppm

Figure 6-2.   The process of biologic magnification of DDT. At each level in the food chain the concentration of DDT is increased and subsequently passed on to the next level. Animals, such as humans and birds, therefore can receive quantities of pesticides concentrated millions of times higher than the amount originally introduced into the environment. (Ross 1977)

species composition of a phytoplankton community, which could have serious effects throughout the food chain. These compounds are found widely dispersed in the animals of the ocean (Table 6-2).

Other organic compounds have also been introduced into the environment in large amounts (Table 6-3). These include 1,2-dichlorethane, used in the production of vinyl chloride and cleaning solvents and as an additive to gasoline and other industrial products; freons, used in aerosol propellants (these have been banned recently in some countries) and refrigerants; and dry cleaning solvents. Most of these compounds have low boiling points and easily enter the atmosphere and, by rain, direct fallout, or ocean–atmosphere interaction, enter the ocean.

Pesticides

Among the more dangerous pesticides in the marine environment are DDT, Aldrin and Dieldrin. They enter the ocean in two main ways—runoff from agricultural areas and via the atmosphere. Most of the publicity has focused on DDT, and its use is presently restricted in many areas of the world. The

Table 6-2. PCB Concentrations from Organisms, Sediments and Water of the Atlantic Ocean[a]

	PCB Concentration (ppb)
Sea mammals	3000
Sea birds	1200
Mixed plankton	200
Finfish from upper ocean waters	50
Finfish from midwater depths	10
Bottom-living invertebrates	1
Deep-sea sediments	1
Sea water	0.001

[a]From Harvey (1974).

amount of DDT in the environment has been estimated to be as high as 1 billion pounds. Most of the DDT enters the ocean via the atmosphere. Once in the ocean it will stay initially in the upper layers of the ocean, but eventually it is carried to deeper layers either by incorporation into organic matter or by sedimentation. These processes are not completely understood. DDT is a fairly stable compound with a half-life of between 10 and 50 years. It will persist in the environment for many years, but studies have indicated that it may not remain this long in the marine environment. DDT sometimes concentrates in organisms at rates higher than that of the normal background concentration, a process known as biologic magnification (Fig.

Table 6-3. Estimated Production and Environmental Leakages of Some Synthetic Organic Chemicals (in millions of tons)[a]

Chemical	General Production		Estimated Leakage to the Environment	
	Annual	Total	Annual	Total
DDT	0.1	2	0.25	1
Aldrin–toxaphene	0.1	1	0.25	0.5
PCB	0.1	1	0.025	0.25
1,2-Dichlorethane	5	2	0.5	
Freons	0.4–0.6		0.4–0.6	
Dry cleaning solvents	2		1–2	
Total synthetic (light organic)	20–30		2–3	
Total synthetic (organic)	100		2	

[a]From U.S. National Academy of Sciences (1971b).

6-2). The process can be very effective; in an exceptional example oysters exposed continuously to waters containing only 0.1 ppb DDT were able to concentrate up to 7.0 ppm of it in their tissue—an increase of 70 000 times within a month.   Recent studies have shown that fortunately biologic magnification does not always work and in some cases even the opposite can be true.   It appears that biologic magnification may be more important among air-breathing organisms (such as seals or birds) than among water-breathing organisms, such as fish.   This appears to be mainly a result of differences in the body chemistry of the organisms.

DDT can have different effects on organisms but generally is not lethal. Its presence in some marine birds results in their having very thin egg shells, which has made successful reproduction difficult.   Some phytoplankton show a significant reduction in photosynthesis rate when DDT is present in the water in concentrations of about 10 ppb; others suffer no adverse effects.

In recent years, with the restrictions on DDT, there has been a corresponding increase in use of other pesticides, such as Aldrin and Dieldrin. These two are even more toxic than DDT and in the United States the Environmental Protection Agency (EPA) is trying to prevent their use.   These actions, although moderately successful, still do not affect the large amounts of these insecticides presently in the environment.   Aldrin and Dieldrin are more soluble in water than DDT and also tend to be concentrated by organisms in their tissues.   These compounds likewise show increased concentrations higher in the food chain and can exceed the background concentration by factors of thousands or more.   Compounds of Aldrin and Dieldrin are toxic to most higher forms of life—more so than DDT, and the danger from these compounds is increased because of their relatively slow rate of decay. Concentrations of a few parts per million can be fatal to some organisms.

Total control of pollution by such pesticides as DDT and others will depend on stopping their use.   However, prohibition would be difficult because they are necessary, in many countries, to control insects carrying diseases, such as malaria, or for crop control.   As yet there is no real substitute for these pesticides.

Excess Nutrients

Near-shore and coastal waters may receive large amounts of industrial and human wastes that contain nutrients, such as phosphorus and especially nitrogen. In the United States as many as 500 000 million tons of nitrogen and 100 000 million tons of phosphorus are discharged each year into its rivers and estuaries.   An oversupply of these nutrients can lead to dramatic increased growth of algae and other phytoplankton (this process is called eutrophication). The growth will discolor the water and make boating, swimming, or other recreational uses of the water unappealing.   If the growth process continues, the decay of the phytoplankton can completely deplete the available oxygen in the water. The reduction or elimination of oxygen would be disastrous to many forms of life and usually leads to mass mortalities—generally among fish. About a decade ago many detergent makers made a considerable reduction in the phosphate content of their products to reduce the possibility of eutrophication. This process generally will not occur in well-mixed waters, where the oxygen content can be maintained by mixing.

Increased nutrient supply can sometimes lead to explosive growth of undesired species. One such type of growth is called "red tide" and seems recently to have become more common. For example, they used to occur along the Florida Gulf Coast about once every 16 years but now occur almost every year. Red tides are caused by a rapid growth of certain dinoflagellate species that sometimes give the water a reddish hue. These phytoplankton are generally not poisonous to the fish, mussels, and clams that feed on them but can be fatal to humans who consume the fish and shellfish. The increased incidence of red tides may be caused by increased supply of nutrients in coastal waters via pollution.

Heavy Metal Pollution

Heavy metals are especially dangerous pollutants that may be introduced into the marine environment by waste and sewage discharge. They are generally present in relatively small concentrations on the order of parts per billion. Dangerous heavy metals include such elements as mercury, cadmium, silver, nickel, and lead (Table 6-4). Some organisms can acquire high concentrations of these metals without apparent harm to themselves. When consumed by humans they then can be extremely dangerous and even poisonous. A recent instance of this concern was the ban on swordfish because of mercury content. A earlier and more severe mercury poisoning event occurred in Minimata Bay, Japan.

The mercury pollution problem at Minimata Bay (called Minimata disease) showed its first signs in the early 1950s. One of the main industries of this small town was the production of certain chemical products including some mercury compounds. As early as 1953, dogs and cats in the area were found dying of some sort of convulsive disease. Similar symptoms were noted among some fishermen of the town, who were found to have nervous system damage. In the beginning investigators examining the problem received little help from the company, which would not divulge what chemical agents were being used or what was being discharged into the sea. Eventually, it was found that the mercury compounds discharged from the factory were being concentrated in shellfish that were, in turn, being eaten by the fishermen. By 1969, over 110 people had contracted this disease and over 45 deaths had resulted; some stillborn children were also found to have symptoms of the disease.

Table 6-4.   World Heavy Metal Production and Potential Ocean Inputs[a]

Element	Mining Production[b] (million tons/year)	Transport by Rivers to Oceans[c] (million tons/year)	Atmosphere Washout[d] (million tons/year)
Lead	3	0.1	0.3
Copper	6	0.25	0.2
Vanadium	0.02	0.03	0.02
Nickel	0.5	0.01	0.03
Chromium	2	0.04	0.02
Tin	0.2	0.002	0.03
Cadmium	0.01	0.0005	0.01
Arsenic	0.06	0.07	—
Mercury	0.009	0.003	0.08
Zinc	5	0.7	—

[a]From U.S. National Academy of Sciences (1971b).
[b]Data from U.S. Department of the Interior.
[c]Bertine and Goldberg (1971).
[d]Estimated from aerosol data of Egorov and others (1970) and Hoffman (1971).

One of the problems concerning heavy metal pollution is how much of these elements humans can tolerate, assuming that they eat an otherwise balanced diet. For example, *Mercenaria mercenaria*, the common quahog, found in many estuaries may have as much as 0.1–0.4 ppm mercury. This value is near, but below, the limit (0.5 ppm) set by the United States for safe human consumption. In theory, therefore, organisms with this amount of mercury are safe to eat. The question remains of how much of such organisms one can eat and what the effect is of such concentrations on human beings. Mercury tends to stay in the human body for long periods of time, having a half-life of about 70 days. This means that toxic amounts can be accumulated by eating small quantities over a period of time. Another aspect is what amount is necessary for marine organisms in their normal diet. Several heavy metals, especially copper and zinc, appear to be necessary for life in the ocean. Unfortunately, as with DDT, there can be an amplification of heavy metals in animal tissue proceeding up the food chain (Fig. 6-2). This explains why such predators as swordfish and tuna, at the top of the food chain, may have heavy metal concentrations high enough to cause concern about their safety for human consumption. The situation is further confused because it is not known how much mercury the organisms may usually have had, prior to our awareness of this problem and to the increased introduction of these metals into the marine environment. For example, the mercury content of museum specimens of tuna, some caught over 93 years ago, was similar to that of recently caught fish (Miller and others 1972).

The input of heavy metals into the marine environment is difficult to evaluate because some are entering the ocean at higher rates by natural activity than by industrial pollution. For example, again considering mercury, uncontrolled input to the ocean by humans is about 4000–5000 tons per year; however, natural weathering of land supplies is about 5000 tons per year, much of which is carried by rivers to the ocean. Moreover, some of the highest mercury values are found in the deep sea near midocean ridges, suggesting still another source—volcanic activity. Obviously, mercury follows several pathways getting to the ocean and, as with many other pollutants, we do not know its oceanwide distribution or understand the important chemical reactions. For example, samples off the European coast have concentrations as low as 16 ppb, whereas concentrations along oceanic ridges can be as high as 400 ppb; higher values occur near industrial areas.

Ocean Dumping

Dumping of material at sea is another major method of pollution. Many countries routinely dump various wastes in the shallow waters off their coasts, assuming that the ocean can easily absorb such ingredients. For example, the annual dumping tonnage for the United States exceeds 45 million tons (Table 6-5), and much may be detrimental to the environment. Some areas are exclusively set aside for dumping purposes, including one about 25 km off New York City.

One argument for dumping waste products at sea is that the material may quickly oxidize and decompose, perhaps even resulting in some benefits to the marine environment. Some recent studies, however, have shown that microbial decay of organic matter in the ocean may proceed at rates 10 to 100 times slower than at similar temperatures on land. This surprising find-

Table 6-5. Ocean Dumping: Types and Amounts, 1968 (in tons)

Waste	Atlantic	Gulf	Pacific	Total	Percent of Total
Dredge spoils	15 808 000	15 300 000	7 320 000	38 428 000	80
Industrial waste	3 013 200	696 000	981 300	4 690 500	10
Sewage sludge	4 477 000	0	0	4 477 000	9
Construction and demolition debris	574 000	0	0	574 000 ⎫	
Solid waste	0	0	26 000	26 000 ⎬	1
Explosives	15 200	0	0	15 200 ⎭	
Total	23 887 400	15 966 000	8 327 300	48 210 700	100

[a]From Council on Environmental Quality (1970) and Ketchum (1973).

ing came from a unique and unplanned experiment, when the research submersible *Alvin* sank in about 1540 m of water. Aboard the submersible when it sank were sandwiches, fruit, and thermos bottles filled with bouillon. When it was recovered about 10 months later the food materials were found to be surprisingly well-preserved. Other experiments in shallower water showed similar results. The implication is obvious—material does not decay as rapidly as thought in the ocean and using the deep sea or the continental shelf as a dumping site for human organic wastes from land may need some serious rethinking.

Among the more dangerous problems of human sewage is that of viruses and the diseases they can cause. It is well known that such sicknesses as hepatitis can be carried via water pathways. In 1972 over 50 000 cases of infectious hepatitis were reported. Although many land or human viruses will not survive in salt water, there can be exceptions created by particular chemical situations where survival of these forms is extended for periods of up to a week or more. Humans coming into direct contact with these waters or into indirect contact by consuming shellfish that have accumulated such viruses can be infected. In comparison to other life forms, relatively little is known about viruses and how they respond in the sewage of the dumping process. No effective and economically feasible method is presently known for removing these organisms from sewage.

During recent years numerous oceanographic expeditions have found large quantities of plastic and polystyrene particles floating on the ocean surface. The concentrations have been as high as 12 000 particles per km^2. The presence of these particles probably reflects increased production of plastics on land and subsequent dumping at sea. The plastic particles can serve as an area of attachment for small plants and animals. Many of the plastics contain polychlorinated byphenyls (PCBs) and thus these particles may be transporting this pollutant to the deep sea.

Some innovative methods have been developed for handling waste products; one technique involves burning the material at sea. This technique is especially appealing in the handling of chlorinated hydrocarbons that can be extremely poisonous. In the past they were either dumped into the sea, burned on land, or converted into other materials, but these disposal methods are only temporarily successful because of the wastes still get into the environment. By combustion the chlorinated organic material can be converted into carbon dioxide, hydrochloric acid, and water—materials that can be fairly well absorbed by the ocean. The scope of the problem is considerable. For example, as many as 400 000 tons of chlorinated hydrocarbons are produced yearly in the Gulf Coast area.

A Dutch vessel, called *Vulcanus* (Fig. 6-3), has been used to safely burn these materials offshore. In Europe these techniques have been used for several years. *Vulcanus* has incinerators that can burn as much as 4200 met-

Figure 6-3. *Vulcanus,* a vessel especially equipped for burning waste products at sea. Its furnances are heated to over 1100°C and then fed waste products at the rate of about 25 tons/h. Preliminary tests indicate that this may be an efficient method for disposing of some toxic pollutants. (Photograph courtesy of the U.S. Environmental Protection Agency)

ric tons of waste and in 1972 it was given permission to burn wastes in the Gulf of Mexico. Tests have shown a 99.9% combustion efficiency, which satisfied government guidelines. In later years the vessel has been used in the Pacific to burn herbicides.

One of the major pollution problems is the safe disposal of waste products from the many nuclear power plants in the world. Some have suggested that such nuclear waste could be safely disposed of in the ocean. This controversial issue is discussed further in Chapter 9, pages 299–301.

United States and Other Legislation Concerning Ocean Dumping

In 1972 the United States Congress passed two pieces of legislation concerning Federal control of water pollution. These acts were the Federal Water Pollution Control Act Amendments (P.L. 92-500) and the Marine Protection, Research and Sanctuaries Act (P.L. 92-532). In the first act discharging of pollutants into the United States territorial sea and contiguous zone was not allowed without a permit. The permits are administrated by the Environmental Protection Agency (EPA) but individual states may also issue permits if their program has been approved by EPA. Conditions for dumping are sometimes ambiguous, balancing a concern for the environment against the apparent need for such dumping and alternative methods of dispersal. The dumping of high-level radioactive wastes and biologic warfare agents was prohibited. Public Law 92-532 (commonly called the Ocean Dumping Act) required permits for transporting and dumping wastes anywhere in ocean waters.

The acts, revised in 1977, hope to attain clean waters for recreational and other uses of the water by the mid-1980s and have zero discharge of pollutants by 1985. All sewage treatment plants will be made to have secondary treatment capabilities which will reduce pollutants entering the sea. Over the first 5 years of the acts' implementation there has been a 90% reduction of ocean-dumped industrial waste.

On the international level, a major conference on dumping was held in London in 1972 and resulted in a Convention on the Prevention of Marine Pollution by Dumping of Wastes and other Matter. This convention prohibits dumping of high-level radioactive wastes and products developed for chemical and biologic warfare. It also prohibits (except where present as trace components) the dumping of mercury and cadmium compounds, organohalogen compounds, persistent plastics and similar synthetic materials, and crude oil and some other petroleum products. Dumping is permitted for such substances as some heavy metals and pesticides, provided special care is taken. United States P.L. 92-532 was subsequently amended to incorporate those parts of the London convention not already in the act.

Hydrocarbon Pollution

Hydrocarbons (oil and gas) are one of the major sources of pollution in the ocean and certainly they attract the most publicity. Hydrocarbons can be introduced into the marine environment by many sources, both natural and unnatural (Table 6-6 and Fig. 6-4). Natural sources include submarine seeps of oil, common, for example, off the California coast, and the decay of marine organic matter, which will produce hydrocarbon material. The unnatural inputs, and the ones that can be controlled, include land-based industrial (including refining) and transportation activities, wrecks, collisions or groundings of oil tankers and other ships, offshore drilling, discharges (deliberate or accidental) from ships, and atmospheric input from the burning of fossil fuels on land. Estimates of atmospheric input of hydrocarbons into the ocean are variable. Nevertheless, recent regulations reducing automobile

Table 6-6.   Petroleum Hydrocarbons in the Ocean

Input per Annum	Million Metric Tons

Transportation
   Tankers, dry docking, terminal
   operation, bilges, accidents
	2.133

Coastal refineries, municipal and industrial waste	0.8
Offshore oil productions	0.08
River and urban runoff	1.9
Atmospheric fallout	0.6
Natural seeps	0.6
Total	6.113

[a]From U.S. National Academy of Sciences (1975).

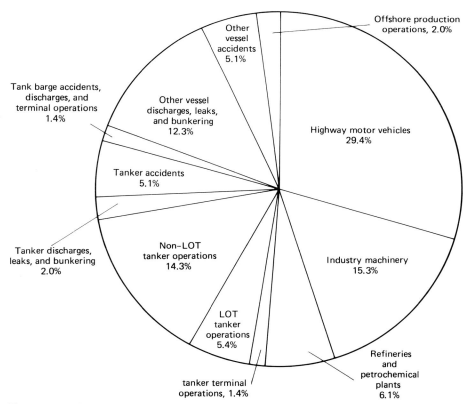

Figure 6-4.   Sources of petroleum pollution in the oceans. LOT indicates load on top (see also Figure 6-11). (Adapted from Parricelli and Keith 1974)

and industrial emissions should aid in reducing this source, although it probably can never be completely eliminated.

The major source of hydrocarbons to the ocean comes from activities
related to transportation of oil. Within United States waters over 8000 spills
occur yearly; however, on a worldwide scale the numbers are even more
dramatic. Over 2.2 billion tons of crude oil are consumed each year
throughout the world, and about half is carried by tankers. There are about
6000 tankers and each year about 6%, or 360, are involved in collisions or
groundings. It is felt that about 0.1% of all oil transported by ships ends up
in the ocean. Estimates of the total amount of hydrocarbons that enter the
ocean vary but generally (see Table 6-6) are at least as high as 6 million tons.
(A ton contains between 6.5 and 7.5 barrels of oil, depending on its weight,
and a barrel of oil contains 42 gallons. Therefore, 6 million tons are about 40
million barrels, or about 1.7 billion gallons.) In losses associated with transportation, the biggest comes from the illegal practice of cleaning oil tanks at
sea by flushing them out with sea water. This problem could be partially
solved by the use of the load-on-top method, which is described later in this
section. Leaking from engines and bilge discharges are other major sources.
Not included in the numbers are major catastrophes, such as the grounding
of the *Torrey Canyon* (Fig. 6-5) in 1967, which lost over 700 000 barrels of

Figure 6-5. The tanker *Torrey Canyon* aground on Seven Stones reef off the
southern coast of England. (Photograph courtesy of Her Majesty's Stationary Office)

Figure 6-6. Oil that ended up in a marina from the offshore oil spill at Santa Barbara. (Photograph courtesy of U.S. Coast Guard)

crude oil. The *Torrey Canyon* was a large ship for that time, but it was in comparison to some more modern tankers, which can be four times its size (Fig. 4-9). Considerable efforts were made to control the spread of the oil, but these were generally ineffective and much of the oil eventually reached the beach and had severe environmental effects. It was estimated that 90% of the marine animal deaths were caused by detergents used to clean up the oil. These detergents caused the oil to form into small drops, which were more easily spread in the marine environment and were therefore more detrimental to marine life. In addition, the detergents themselves are a pollutant.

Another well-publicized major oil spill occurred in 1969 off Santa Barbara, California. In this instance, a high-pressure offshore oil well blew out, resulting in over 700 000 gallons (about 16 000 barrels) of oil being discharged into the environment. Much of this oil drifted ashore (Fig. 6-6), coating beaches and covering many sea birds that dove unknowingly into the oil-covered water. Oil leaked into the area for over 300 days and caused several millions of dollars damage. There has been considerable debate about the environmental effects of this spill, in large part because adequate background data are lacking; for example, what were the quantity and quality of life in this area before the spill? The offshore Santa Barbara area is under considerable pressure by oil pollution because the natural oil seepage in the area may be as high as 50 to 75 barrels per day. At this rate it would take a little over 222 days to equal the amount spilled from the blown well. The effects are obviously different in part because the well discharged its oil into one area, whereas the seeps are widely spread. Natural oil and gas seeps are common occurrences in some areas and may contribute a modest amount of the oil in the environment (Table 6-6). Nevertheless, the presence of natural seeps does not detract from the importance of the Santa Barbara spill, although it does show some of the problems in evaluating its impact.

A small, although important, oil spill occurred in 1968 when a Liberian registered tanker *Ocean Eagle* went aground near the entrance to San Juan Harbor, Puerto Rico (Fig. 6-7). The vessel split in half during heavy seas, but

Figure 6-7. Both halves of the tanker *Ocean Eagle* lying aground off the mouth of San Juan Harbor, Puerto Rico (seen at top of figure). Oil spilled from the hull for 2 weeks, with much of it reaching the beach area. (Photograph courtesy of U.S. Coast Guard)

Figure 6-8. The 1067-ft (325-m) tanker *Metula* lying around in the Strait of Magellan near the coast of Chile. A tug is trying to assist the ship. The *Metula* lost over 300 000 barrels of oil. (Photograph courtesy of U.S. Coast Guard)

its 33 crew members were saved. Eventually more than 2 million gallons of oil were discharged, which resulted in numerous resort hotels having to close their beaches during the height of the tourist season. A much larger spill occurred in August of 1974, when the *Metula,* a Dutch supertanker, ran aground 2 miles north of Tierra del Fuego at the extreme end of South America (Fig. 6-8). The tanker was carrying 1.58 million barrels of oil but fortunately lost less than half of its cargo. About 25% of the oil reached Chile's shores and covered about 75 miles of beach. Estimates are that at least 40 000 birds were killed and that it would take 10 years for the environment to return to normal conditions.

One of the best studied oil spills occurred in 1969 when the oil barge, Florida, ran aground in Buzzards Bay, Massachusetts, off West Falmouth. The grounding caused about 700 tons of No. 2 fuel oil to be discharged into the coastal waters. Onshore winds and tides carried much of it into the very productive marshlands, resulting in extensive damage. The area has been studied for over a decade and some of the affected areas have yet to recover. In some localities the oil penetrated as much as 2 ft into the sediment. The shellfish of the area were especially affected by the oil, and total ecologic damage in terms of local fishing and shellfish loss may eventually exceed $1 million. It should be emphasized that the Falmouth spill was a relatively small one. What distinguished it was the fact that it was studied in detail over a period of many years, and that these studies showed how long lasting pollution effects can be.

An amazing sequence of oil spills occurred in late 1976 and early 1977. These included:

15 December 1976. Liberian registered[1] *Argo Merchant* went aground off Nantucket and released 7.6 million gallons of oil.

17 December 1976. Liberian registered *Sansinena* exploded in Los Angeles harbor, killing nine persons.

24 December 1976. Liberian registered *Oswego Peace* spilled 2000 gallons of oil in the Thames River, Connecticut.

27 December 1976. Liberian registered *Olympic Games* ran aground in Delaware River and spilled 133 500 gallons of oil.

30 December 1976. Panamanian registered *Grand Zenith* disappeared off Nova Scotia carrying 8.2 million gallons of oil.

4 January 1977. Liberian tanker *Universe Leader* ran aground in Delaware River but there was no spill.

5 January 1977. United States registered *Austin* spilled 2100 gallons of oil into San Francisco Bay.

7 January 1977. Liberian tanker *Barbola* carrying 13 million gallons of crude oil, ran aground off Texas but refloated with no loss of oil.

The record of the largest oil spill seems to belong to the Liberian registered *Amoco Cadiz,* which broke up off the Brittany coastline in March of 1978. It was carrying 68 million gallons of oil, much of which ended up on the nearby beaches. The immediate effect of the spill was the severe damage to many local coastal shellfisheries, as well as pollution of a large stretch of the coastal zone. An international group of scientists is in the process of assessing the total effects.

Ironically, until recently public opinion was rarely raised against tankers but was almost always directed against offshore drilling. Drilling for oil is not without its potential for pollution but is considerably safer than shipping. For example, in the Gulf of Mexico area, where there are over 10 000 offshore wells, the spillage in the interval 1971–1975 was 51 421 barrels of oil (the *Argo Merchant* spilled over three times this amount). According to the U.S. Geological Survey most of this came from five incidents, which accounted for 85.5% of the total. It was estimated that for each 35 219 barrels produced, one was spilled for a rate of 0.0028%—considerably better than the 0.1% rate from tankers. Perhaps of secondary importance is that the wells are generally owned by responsible companies, so liability can be established and imposed, whereas this can be very difficult for tankers (see page 73). Pollution from oil platforms and drilling is not without its dangers

[1]See Chapter 4 concerning registry of ships.

especially because of chronic, low-level (small amounts over long periods of time) input of hydrocarbons. The effects of this type of pollution are often not very obvious but can cause significant damage to the local ecology.

Hydrocarbons are composed of literally thousands of different organic compounds and are an extremely complex material. All of the compounds contain hydrogen and carbon and some have oxygen, nitrogen, or sulfur. The products that can be obtained from crude oil are as diverse as asphalt and gasoline, which gives an indication of its range in chemical composition. It is probably true that no one sample of oil has ever been completely analyzed. However, although crude oils from different fields will have different compositions and physical properties, such as viscosity and specific gravity, they will contain many similar compounds.

When oil enters the marine environment, several things will happen. Some of the lighter and more volatile compounds will evaporate into the atmosphere. Some of the components will dissolve into the water, some will sink, some will decay under biologic weathering, some will be absorbed into the food chain if consumed by filter feeders, and some may enter into the bottom sediments (Fig. 6-9). The different processes will be affected by the wind (an aid in dispersal), by temperature (with colder temperature the oil

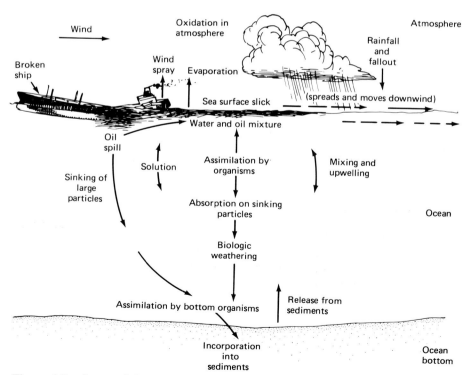

Figure 6-9. Some of the possible pathways that an oil spill may travel.

may form larger lumps; also cold will retard biologic degradation), as well as by the original composition of the oil (light or heavy oil).   Bacteria can degrade or metabolize the oil but rarely at a rate able to prevent major damage.

The oil that sinks will persist in the environment for a considerable time, because besides being unable to evaporate, the deeper waters have lower temperatures and oxygen contents that reduce the possibility of biologic degradation and oxidation of the oil. Once in the sediment, biologic degradation may cease if the sediments are anaerobic, which they generally are.

Hydrocarbons are extremely toxic and poisonous to essentially all marine organisms.   The toxic effects can remain for years, especially if the oil trapped within the bottom sediment is slowly released.   After a pollution event it will usually take a considerable period of time until the fauna and flora return to their original state.   Low-level amounts of oil in the environment can affect biologic processes, such as reproduction and feeding.   These processes are influenced by chemical compounds that are present in extremely small quantities.   The oil may either block these chemical messengers or create new ones that are confusing to organisms.

Once an oil spill occurs there are several types of measures that can be implemented to reduce or restrict its damage.   Unfortunately, no really adequate technique exists for handling oil spills in rough water.   The basic techniques used are:

Burning or combustion

Sinking of the oil

Mechanical removal or containment

Biologic degradation

In some early spills, detergents were used as disperants, but they were found to be as toxic as the oil.   Some nontoxic dispersants have been developed, but generally they still are dangerous.   The objective of the disperants is to reduce the surface tension of the oil so that it forms small drops.   The effect is perhaps satisfying from an esthetic viewpoint, but it is environmentally dangerous.   The smaller size of the particles permits more rapid solution of toxic oil compounds into the water.   The concentration of these compounds, by this method, is higher than it would be by normal dispersion.

Burning or combustion probably is not a bad method for treating an oil spill, but generally it is very difficult to do, unless done immediately.   If successful, it will lead to some atmospheric pollution.   Sinking of the oil can also be esthetically pleasing but, as previously mentioned, produces a more damaging situation.

Mechanical removal or containment of the spill is an ideal method of treatment, and numerous techniques exist.   In some port areas, ships that are being fueled will be surrounded by containment devices—such as floating booms completely encircling the ship. Special ships have been designed that can skim up oil on the water and collect it. Unfortunately, few of these devices can operate effectively in even moderately rough seas.

Biologic degradation using marine bacteria will work but there are many unknowns. It has been hoped that some strain of bacteria could be developed that would rapidly digest oil to acceptable by-products. These bacteria have not yet been developed and the present forms can produce dangerous or toxic intermediate products. The seeding of an oil spill with bacteria may also be a difficult operation. In addition, oxidation of oil requires a large amount of oxygen—the oxygen in over 300 000 gallons of sea water is needed to oxidize just 1 gallon of oil.

When the *Argo Merchant* sank, the U.S. National Oceanic and Atmospheric Administration (NOAA) made an experiment. They dropped an estimated 12 000 floating cards into the seas in the area of the disaster (Fig. 6-10). The objective was that these cards, which were plastic, would float and drift in the same direction as the oil and water and when subsequently found they could tell something about the movement of the oil and water.

Figure 6-10. An aerial photograph, from 5500 ft (1676 m) of the *Argo Merchant*. The oil is clearly visible. (Taken in December of 1976 by the U.S. National Aeronautics and Space Administration)

About half the cards were dropped into the waters between the oil spill and the islands of Nantucket, Martha's Vineyard, and Long Island. The other 6000 or so floaters were dropped into the oil itself to directly measure its movement and dispersal activity. Amazingly, not one of these cards has been returned to NOAA—a point that indicates how little we really know about some of the processes active in the ocean.

The effects of oil pollution can be either of an acute or a chronic nature. An acute effect is one that results from a single spill or discharge. The major result will be the death of numerous organisms but, depending upon the area and the oceanographic conditions, the fauna and flora can eventually recover. Chronic effects occur when the oil pollution either occurs continuously or without sufficient time for the area to recover. Often, the impact of the chronic effect is more long lasting than that of single inputs.

One way to reduce oil pollution by tankers is by use of the load-on-top system (Fig. 6-11). Tankers, after they discharge their oil, must add some water as ballast to their tanks; otherwise they will ride too high when at sea. This water will mix with the oil clinging to the walls of the tank. If the seas are calm the oil and water mixture will start to separate (oil on top) after a

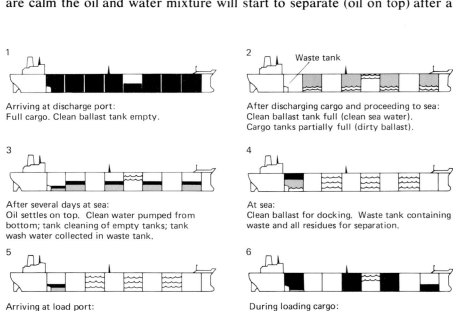

1
Arriving at discharge port:
Full cargo. Clean ballast tank empty.

2 Waste tank
After discharging cargo and proceeding to sea:
Clean ballast tank full (clean sea water).
Cargo tanks partially full (dirty ballast).

3
After several days at sea:
Oil settles on top. Clean water pumped from bottom; tank cleaning of empty tanks; tank wash water collected in waste tank.

4
At sea:
Clean ballast for docking. Waste tank containing waste and all residues for separation.

5
Arriving at load port:
Clean ballast for docking. Waste tank drained of all clean water, leaving only collected residue. Before loading, all clean water pumped into sea.

6
During loading cargo:
Waste tank loaded on top of residues.

[☷] Clean seawater   [■] Crude oil   [▨] Oil contaminated seawater

Figure 6-11. Operation of the load-on-top technique. (Adapted from U.S. Environmental Protection Agency 1975)

few days. With many ships the oil and water mixture will be pumped into the ocean prior to the ship's receiving a new cargo. However, if the water were to be pumped out from below and the upper oil–water mixture saved and collected in one tank, considerable pollution could be prevented (Fig. 6-11). This procedure has been recommended by the Intergovernmental Maritime Consultative Organization (IMCO) for adoption by participating countries. It would, however, only apply to new ships that are larger than 70 000 dwt tons. At present, the required number of countries, including the United States, have not signed the agreement.

The question of oil pollution is a highly emotional one, but some important facts should be considered. For example, according to a U.S. National Academy of Science report in 1975, *Petroleum in the Marine Environment*, the amount of oil entering the total marine environment (see Table 6-6) from offshore drilling (80 000 tons per year) is 7.5 times less than from natural seeps (600 000 tons per year) and 26 times less than that from transportation (2 133 000 tons per year). Most public concern, as previously mentioned, seems to be centered on offshore drilling rather than on oil pollution from tankers. A more realistic appraisal might be what is the best way, ecologically, to get the needed oil. For example, the potential drilling along the east coast of the United States has drawn considerable opposition from people concerned about the environment. However, offshore drilling in the United States has been relatively safe. From 1964 to 1971 there were 16 major spills from 10 234 producing wells. The oil released in the 8 years was about 46 000 tons; this number, although it could be lowered by stronger government regulation, is still considerably less than the 200 000 or so tons that the *Metula* carried (Fig. 6-8). An offshore drilling program with buried pipelines (a relatively safe way of transporting oil) and in-shore refineries would, in many instances, be several times safer ecologically than would bringing in the oil by tankers, of which about 6% per year have accidents. Unfortunately, what sometimes happens with offshore platforms is that the oil, instead of being moved in pipelines, is put on tankers and transported to a refinery. Under this scenario the potential for pollution to the environment is increased considerably. Part of the concern for offshore drilling platforms, besides pollution, can also come from concern about interference with other activities (fishing, for example), visual aspects, and onshore activities associated with drilling, such as storage and transportation facilities.

Certainly a major problem is how drilling and drilling platforms interfere with fish and fishing. Unquestionably, there is a problem when a spill occurs, but there may be some beneficial effects from drilling platforms. The platforms can provide protection and a habitat to the smaller animals in the food chain. In this manner certain larger species will be attracted to the area. As an example of this, since the first operation of oil platforms in the Gulf of Mexico in 1947, the fishing catch has risen from 300 million pounds per year to 1.62 billion pounds per year in the area. Part of this increase is undoubtedly caused by the protection and habitat offered by the platforms. The exact cause and effect relationship however, is hard to ascertain because there also has been an increase in fishing effort during this period of time.

Pollution Resulting from Exploration and Exploitation of Marine Mineral Resources

The mineral resources of the ocean, besides oil and gas mentioned in the previous section, also include sand and gravel on the continental shelf and manganese nodules in the deep sea (see Chapt. 5).

Offshore mining of sand and gravel can have a deleterious effect on the marine environment (Fig. 6-12). The dredging or pumping operation needed to raise the material from the bottom will destroy bottom organisms in most instances. This in turn will damage or change the entire ecosystem as well as cause alteration or modification of spawning areas. The operation will also introduce large amounts of fine-grained material into the water column, producing turbidity that can have a damaging effect on the plankton, and on the benthos when the material settles to the bottom. Another problem could result if the bottom sediments were to contain pollutants or toxic materials: These materials would be then reintroduced and resuspended into the water column with possible harmful effects. Finally, once the dredging operation is completed, the bottom topography will have been modified, in some instances enough to alter the previous current regime or perhaps to make bottom trawling by fishermen difficult or impossible. The removal of large

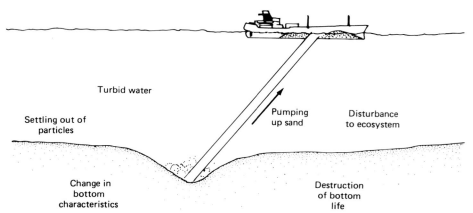

Figure 6-12. Some possible environmental damage caused by a sand dredging operation.

Figure 6-13. Aerial photograph of dye injected into the mining effluent brought up to the surface from a manganese nodule mining operation. The photograph shows the spreading 8 min after the discharge. (See Amos and others 1972, for a further discussion.) (Photograph courtesy Dr. A.F. Amos, University of Texas)

amounts of sand from near-shore regions will effect the sedimentary cycle of the adjacent beaches (see Chapt. 8) and could lead to rapid beach erosion.

It is difficult to predict environmental problems prior to dredging, without some preliminary study or modeling. Long-term observations of currents, beach changes, fauna, and the like would be very helpful. As discussed in Chapter 5, the economics of offshore dredging are such that the shallow areas close to land are the most desireable; these are also the most ecologically vulnerable.

Mining of manganese nodules from the deep sea (Fig. 5-16—5-23) presents many interesting oceanographic, technologic, and pollution problems. The water that will come to the surface with the nodules will be from considerable depth and so will be colder and have a slightly different salinity and probably higher nutrient content than the surface waters. It will also contain large amounts of mud from the bottom (Fig. 6-13). The environmental effects of the mining operation have good and bad aspects. If the deeper water spreads out on the surface, its relatively high nutrient content may be beneficial for phytoplankton growth; however, the large amount of suspended matter (mud) will reduce the transparency of the water and therefore reduce light penetration depth, and thus the water depth available for photosynthesis. The suspended material may also take years before it settles back to the bottom, and in doing so it can alter the chemistry of the water.

The U.S. National Oceanic and Atmospheric Administration made a 3-year study to ascertain the effect of mining manganese nodules; the program is called DOMES (Deep Ocean Mining Environmental Study). Their reports have shown that the material brought from the ocean bottom by a mining operation could reduce light penetration in upper waters and this, in turn, would have an effect on photosynthesis and productivity. Generally, however, this would only be temporary, with the water clearing to about half its normal level within 24 hours. Changes in water temperature, salinity, oxygen, and nutrients would probably be quite small for the mining operation. It was felt, however, that continued monitoring of mining tests is necessary to understand all the environmental effects.

The program has estimated that mining ships could recover as much as 5000–10 000 metric tons of nodules a day. Such ships would probably operate around the clock for about 300 days during the year. In the process of recovering 5000 tons of nodules, the DOMES scientists estimated that a mining system would have to take the nodules from a 222-acre site, which is about 0.9 km². In the process of bringing nodules up to the surface the system would probably also suck up about 36 000 tons of sediment, 1727 lb of living organisms, and about $3\frac{1}{2}$ million cubic feet of water. Much of the water could be put back near the bottom. At the surface about 1000 metric tons of sediment would be discharged, including an additional 50 metric tons of broken up nodules and about 51 lb of biota and about 700 000 ft³ of water.

The effect of a manganese nodule mining operation on the ocean bottom is certainly more difficult to ascertain. In general, the characteristics and ecologic parameters of the deep-water biologic population are not well known, but surely they could be adversely affected by the mining operation. The falling suspended matter from the water carried to the surface could bury some benthic organisms. However, the number of organisms in most manganese nodule areas is generally very small (many consider these areas to be biologic deserts) and therefore any disturbance may have little importance in the overall biologic system of the ocean. Alternatively, it must be emphasized that little is known about this part of the ocean and the long-range effects of mining could have unanticipated results. A final point is that the mining of copper, nickel, cobalt, and manganese (important elements in manganese nodules) from mines on land in generally a very destructive operation and in all probability is several times more environmentally damaging than a marine operation.

Radioactive and Thermal Wastes

The input of radioactive material into the ocean from past nuclear explosions does not appear to have produced major problems and is less than that from naturally occurring radioisotopes. This comparison does not tell the entire story, because different radionuclides will have different toxicities and the naturally occurring ones are generally less dangerous than the ones from nuclear bombs. The treaty on the prohibition of emplacement of nuclear weapons on the sea bed (which only refers to areas outside the territorial sea and contiguous zone—see Chapt. 3) and the general ban on atmospheric testing of nuclear bombs signed by most nuclear powers should keep this type of pollution to a minimum in the future. In addition, there are fairly strict international controls concerning dumping of radioactive waste from nuclear plants into the ocean. It also appears that the artificial radioactive substances presently introduced into the ocean by atomic bomb blasts exceeds the oceanic input from land-based reactions or nuclear fuel processing plants. The estimated bomb imput is about 1/1000th of the total natural radioactivity in the ocean. The major danger of radioactive material is not of polluting the whole ocean but of damaging a small, restricted area where the material cannot be rapidly dispersed or diluted. With proper controls and regulations this possibility could be kept to a low probability.

A major possibility of radioactive pollution arises if the large amount of spent nuclear fuel is put into deep-sea beds as its "final" resting place. This is discussed further in Chapter 9, page 299.

A relatively new problem for the ocean, and especially for coastal areas, is that of thermal pollution or thermal waste. Thermal pollution results from the large volumes of water needed to cool electrical or nuclear power plants. Many of these plants are sited near a source of water for cooling, such as estuaries, lakes, or coastal areas. The water, after being used for cooling, is generally directly returned to the environment but it can be 10–20°C higher in temperature than it was initially. This change can be extremely harmful or fatal to most marine fauna and flora. In some instances, holding ponds or cooling towers permit the water to return to a more normal temperature before it is returned to the environment. In other situations the warm or hot water can be used in a beneficial situation, such as in an aquaculture program to increase growth of oysters or lobsters.

There are two main adverse biologic results from thermal pollution: It affects the metabolic activities of the fauna and it decreases oxygen solubility. Another possible problem of using water for cooling is that generally a very large volume is needed. This involves complex and large pumping systems which can suck in small marine life, especially plankton, often killing them. With the anticipated increase in nuclear power plants in the future there will be increased demands on using water for cooling systems.

Conclusions

There are numerous varieties of pollutants in the ocean, especially in the coastal zone. In many instances, adequate data concerning pollution input or conditions prior to the Industrial Revolution are not available. It is probable that some forms of marine pollution will continue to get worse, even if controls are established. The curtailment or reduction of many forms of pollution will require international cooperation and large-scale financing. Many developing countries view pollution as a problem of the highly industrialized countries, because they are the ones that cause most of it, and so are unwilling to contribute to its solution. The Law of the Sea Conferences and other international meetings have developed useful procedures to regulate and reduce some forms of pollution. Within the United States some recently enacted legislation has reduced the incidence and effects of offshore dumping.

In the marine environment, pollution can have three basic effects (any or all can happen):

1. It can directly destroy the organisms within the polluted area.
2. It can alter the physical and chemical properties of the environment, thus favoring or excluding specific organisms.
3. It can introduce substances that are dangerous to higher forms of life, such as human beings, but that are relatively harmless to lower forms of life.

Probably the most critical aspect of marine pollution is the large quantity of pollutants that enters the coastal zone and estuaries. For example, over 30 billion gallons of industrial and municipal wastes are discharged daily across the United States into the coastal zone. In many instances this input exceeds the capacity of these waters to cleanse themselves, and even with proper treatment and management large amounts of potentially damaging pollutants remain in the bottom sediments.

The open ocean also has its share of detectable pollutants. However, its size and capacity make it less an immediate problem than our near-shore waters. As emphasized many times before, these near-shore areas are the localities of greatest marine use, both by humans and the organisms of the sea (see Chapt. 8 for more details).

There are several ways that the pollution problem can be reduced and clearly one of the best methods is to control it at its source. Certainly a major method is to recycle as many materials as we can, instead of disposing of them after only a single use. There are other more innovative techniques that could actually even produce a benefit. For example, many of the waste products, especially those with high organic or nutrient content, and thermal effluents could be used to increase biologic productivity via aquaculture or mariculture techniques. These ideas are presently being implemented in scale models and offer considerable potential for the future. It is also possible to produce proteins from petroleum wastes by using microorganisms and yeast and other products from industrial pollution. Sludge materials, if properly treated, can be used for soil additives and fertilizers.

Within the ocean itself there are many situations in which the harmful aspects of a substance are not clear or it is not known how much material the ocean can tolerate. For many naturally occurring pollutants it is even possible to show beneficial effects. These are most obvious in open areas having good circulation. As an indication of the complexity of the problem and the difficulty in accepting some pollution, consider the point made by John Isaacs (cited in Bascom 1974). He noted that the 6 million metric tons of anchovies off southern California produce, without apparent harm to the environment, as much fecal material as 90 million people (10 times the population of Los Angeles) and that the anchovies only comprise one of the many hundreds of species of marine life in the region. Because human fecal material is similar to that of marine animals, the implications are obvious.

References

Ahern, W.R., 1973, *Oil and the Outer Coastal Shelf: The Georges Bank Case.* Cambridge, Mass.: Ballinger.

Allen, A.A., and R.S. Schleuter, 1970, Natural oil seepage at Coal Oil Point, Santa Barbara, California: *Science,* v. 170, p. 974–977.

*Amos, A.F., C. Garside, K.C. Haines, and O.A. Roels, 1972, Effects of surface discharged deep-sea mining effluent: *Marine Technol. Soc. Jour.,* v. 6, n. 4, p. 40–45.

Amos, A.F., O.A. Roels, and A.Z. Paul, 1976, Environmental Baseline Conditions in a Manganese-Nodule Province in April–May 1975. Paper presented at Eighth Annual Offshore Technology Conference, Houston, Texas, 3–6 May 1976.

Atema, J., 1977, The effects of oil on lobsters: *Oceanus,* v. 20, n. 4, p. 67–&(.

Ayers, R.C. Jr., H.O. Jahns, and J.L. Glaeser, 1974, Oil spills in the Arctic Ocean: Extent of spreading and possibility of large scale thermal effects: *Science,* v. 186, p. 843–846.

Barber, R.T., and A. Vijayakumar, 1972, Mercury concentrations in recent and ninety-year-old benthopelagic fish: *Science,* v. 178, p. 636–638.

*Bascom, W., 1974, The disposal of waste in the ocean: *Sci. Amer.,* v. 231, n. 2, p. 16–25.

Bertine, K.K., and E.E. Goldberg, 1971, Fossil fuel combustion and the major sedimentary cycle: *Science,* v. 173, p. 233–235.

Blumer, M., 1971, Scientific aspects of the oil spill problem: *Environ. Affairs,* v. 1, n. 1, p. 54–73.

*References indicated with an asterisk are cited in the text; others are of a general nature.

Blumer, M., 1972a, Submarine seeps: Are they a major source of open ocean oil pollution? *Science*, v. 176, p. 1257–1258.

Blumer, M., 1972b, Oil contamination and the living resources of the sea: *In:* Ruivo, M. Ed., *Marine Pollution and Sea Life.* Published by arrangement with the Food and Agricultural Organization of the United Nations by Fishing News (Books) Ltd., p. 476–481.

Blumer, M., and J. Sass, 1972, Oil pollution: Persistence and degradation of spilled fuel oil: *Science*, v. 176, p. 1120–1122.

Blumer, M., P.C. Blokker, E.B. Cowell, D.F. Duckworth, 1972, Petroleum: *In* E.G. Goldberg, Ed., *A Guide to Marine Polluton.* New York: Gordon and Breach, p. 19–40.

Blumer, M., H.L. Sanders, J.F. Grassle, and G.R. Hampson, 1971, A small oil spill: *Environment*, v. 13, n. 2, p. 2–12.

Booda, L., 1976, Marine mining faces bright economic future: *Sea Technol.*, August, p. 23–28.

Bowen, V.T., 1974, Transuranic elements and nuclear wastes: *Oceanus*, v. 18, n. 1, p. 42–54.

Brook, A.J., and A.L. Baker, 1972, Chlorination at power plants: Impact on phytoplankton productivity: *Science*, v. 176, p. 1414–1415.

Brown, R.A., and H.L. Huffman Jr., 1976, Hydrocarbons in open ocean waters: *Science*, v. 191, p. 847–849.

Butler, P.A., 1969, Pesticides in the sea: *In Encyclopedia of Marine Resources.* New York: Van Nostrand Reinhold, p. 513–516.

Carpenter, E.J., 1974, Power plant entrainment of aquatic organisms: *Oceanus*, v. 18, n. 1, p. 34–41.

Carpenter, E.J., and K.L. Smith, Jr., 1972, Plastics on the Sargasso Sea surface: *Science*, v. 175, p. 1240–1241.

Carpenter, E.J., S.J. Anderson, G.R. Harvey, H.P. Miklas, and B.B. Peck, 1972, Polystyrene spherules in coastal waters: *Science*, v. 178, p. 749–750.

Clark, J.R., 1969, Thermal pollution and aquatic life: *Sci. Amer.*, v. 220, n. 3, p. 19–27.

Colton, J.B., F.D. Knapp, and B.R. Burns, 1974, Plastic particles in surface waters of the northwestern Atlantic: *Science*, v. 185, p. 491–497.

Council on Environmental Quality, 1970, *Ocean Dumping, A National Policy.* A Report to the President prepared by the CEQ. 45 pp.

Dahle, E. Jr., 1973, *The Continental Shelf Lands of the United States: Mineral Resources and the Laws Affecting Their Development, Exploitation and Investment Potential.* University of North Carolina Sea Grant Pub. No. UNC-SG-73-11, Chapel Hill, N.C., 50 pp.

Danzig, A.L., 1973, Marine pollution–A framework for international control: *Ocean Management*, v. 1, p. 347–379.

Davis, C.C., 1972, The effects of pollutants on the reproduction of marine organisms: *In:* Ruivo, M., Ed., *Marine Pollution and Sea Life.* Published by arrangement with the Food and Agricultural Organization of the United Nations by Fishing News (Books) Ltd., p. 305–311.

Devanney, J.W., V. Livanos, and J. Patell, 1971, Economic Aspects of Solid Waste Disposal at Sea. Massacussetts Institute of Technology Sea Grant Pub. N. 71-2, Cambridge, Mas., 113 pp.

Duce, R.A., and J.G. Quinn, 1972, Enrichment of heavy metals and organic compounds in the surface microlayer of Narragansett Bay, Rhode Island: *Science*, v.
176, p. 161–163.

Duce, R.A., P.L. Parker and C.S. Giam, 1974, Pollutant Transfer to the Marine Environment. Deliberations and Recommendations of the NSF/IDOE Pollutant
Transfer Workshop held in Port Aransas, Texas, 11–12 January 1974, 55 pp.

Dyrssen, D., and D. Jagner, Eds., 1972, *The Changing Chemistry of the Oceans,
Proceedings of the Twentieth Nobel Symposium.* New York: John Wiley &
Sons.

Economic Aspects of Solid Waste Disposal at Sea. 1970. A report prepared for the
National Council on Marine Resources and Engineering Development, Mas-
sachusets Institute of Technology, Cambridge, Mass., 124 pp.

*Egorov, V.V., T.N. Zhigalovskaya, and S.G. Malakhov, 1970, Microelement con-
tent of surface air above the continent: *Jour. Geophys. Res.*, v. 75, p. 3650–3656.

Elder, D.L., and S.W. Fowler, 1977, Polychlorinated biphenyls: Penetration into the
deep ocean by zooplankton fecal pellet transport: *Science,* v. 197, p. 459–461.

English, T.S., 1973, *Ocean Resources and Public Policy.* Seattle: Univ. Washington
Press.

Farrington, J.W., 1977, The biogeochemistry of oil in the ocean: *Oceanus,* v. 20, n. 4,
p. 5–14.

Gerard, R., 1976, Environmental effects of deep-sea mining: *Marine Technology Society Jour.,* v. 10, n. 7, p. 7–16.

Gissespie, R.J., and C.R. Phillips, 1976, Treatment of oily wastes from ships: *Marine
Technology Society Jour.,* v. 10, n. 5, p. 19–26.

Goldberg, E.D., 1972, Man's role in the major sedimentary cycle: *In: The Changing
Chemistry of the Oceans.* New York: John Wiley & Sons, p. 267–288.

Goldberg, E.D., 1974, Marine pollution: Action and reaction times: *Oceanus* v. 18,
n. 1, p. 6–16.

Goldberg, E.D., 1976, *The Health of the Oceans.* Paris: UNESCO Press.

Gross, M.G., 1972a, Geologic aspects of waste solids and marine waste deposits,
New York metropolitan region: *Geol. Soc. Amer. Bull.,* v. 83, p. 3163–3176.

Gross, M.G., 1972b, Waste removal and recycling by sedimentary processes: *In:
Ruivo, M. Ed., Marine Pollution and Sea Life.* Published by arrangement with
the Food and Agricultural Organization of the United Nations by Fishing News
(Books) Ltd., p. 152–158.

Gulland, J.A., 1974, *The Management of Marine Fisheries.* Seattle: Univ. Washington Press.

Halstead, B.W., 1972, Toxicity of Marine Organisms caused by Pollutants: *In:
Ruivo, M., Ed., Marine Pollution and Sea Life.* Published by arrangement with
the Food and Agricultural Organization of the United Nations by Fishing News
(Books) Ltd., p. 584–594.

Hardy, M., 1973, Offshore development and marine pollution: *Ocean Devel. Internatl. Law Jour.,* v. 1, n. 3, p. 239–273.

*Harvey, G.R., 1974, DDT and PCB in the Atlantic: *Oceanus,* v. 18, n. 1, p. 18–23.

Harvey, G.R., W.G. Steinhauer, and J.M. Teal, 1973, Polychlorobiphenyls in North
Atlantic ocean water: *Science,* v. 180, p. 643–644.

Hobbs, P.V., H. Harrison, and E. Robinson, 1974, Atmospheric effects of pollutants:
Science, v. 183, p. 909–915.

*Hoffman, G.L., 1971, Trace Metals in the Hawaiian Marine Atmosphere. Ph.D. Thesis, University of Hawaii, Honolulu.

Hom, W., R.W. Risebrough, A. Soutar, and D.R. Young, 1974, Deposition of DDE and polychlorinated biphenyls in dated sediments of the Santa Barbara Basin: *Science*, v. 184, p. 1197–1199.

Hood, D., 1971, *Impingement of Man on the Oceans.* New York, Wiley-Interscience; 738 pp.

How hot is ocean incineration? 1975. *Environ Science & Technology,* v. 9, n. 5, p. 412–413.

Inman, D.L., and B.M. Brush, 1973, The coastal challenge: *Science,* c. 181, p. 20–31.

Jannasch, H.W., and C.O. Wirsen, 1973, Deep-sea microorganisms: In situ response to nutrient enrichment: *Science,* v. 180, p. 641–643.

Jannasch, H.W., K. Eimhjellen, C.O. Wirsen, and A. Farmanfarmaian, 1971, Microbial degradation of organic matter in the deep sea: *Science,* v. 171, p. 672–675.

*Ketchum, B., 1973, A realistic look at ocean pollution: *Marine Technology Society Jour.,* v. 7, n. 7, p. 8–15.

Knebel, H.J., 1974, Movement and Effects of Spilled Oil Over the Outer Continental Shelf—Inadequacy of Existent Data For the Baltimore Canyon Trough Area. U.S. Geol. Survey Circular 702, Washington, D.C., 17 pp.

Krebs, C.T., and K.A. Burns, 1977, Long-term effects of an oil spill on populations of the salt-marsh crab *Uca pugnax: Science,* v. 197, p. 484–487.

Kullenberg, G.E.B., 1974–1975, Ocean dumping sites: *Ocean Management,* v. 2, p. 189–209.

Longwell, A.C., 1977, A genetic look at fish eggs and oil: *Oceanus,* v. 20, n. 4, p. 46–58.

Lumsdaine, J.A., 1976, Ocean dumping regulation: An overview: *Ecol. Law Quart.* v. 5, n. 4, p. 753–792.

McKelvey, V.E., J.I. Tracey Jr., G.E. Stoertz, and J.G. Veder, 1969, *Subsea Mineral Resources and Problems Related to Their Development.* U.S. Geol. Survey, Washington, D.C., 26 pp.

Mercury in the Environment, 1970, Geol. Survey Prof. Paper 713, U.S. Government Printing Office, Washington, D.C., 65 pp.

Mercury in the Environment: Natural and Human Factors: *Science,* v. 171, p. 788–789.

Milgram, J., 1977, The cleanup of oil spills from unprotected waters: Technology and policy: *Oceanus,* v. 20, n. 4, p. 86–93.

Miller, D.S., D.B. Peakall, and W.B. Kinter, 1978, Ingestion of crude oil: Sublethal effects in herring gull chicks: *Science,* v. 199, p. 315–317.

*Miller, G.E., P.M. Grant, R. Kishore, F.J. Steinkruger, F.S. Rowland, and V.P. Guinn, 1972, Mercury concentrations in museum specimens of tuna and swordfish: *Science,* v. 175, p. 1121–1122.

Moore, G., 1972, The control of marine pollution and the protection of living resources of the sea: *In:* Ruivo, M., Ed., *Marine Pollution and Sea Life.* Published by arrangement with the Food and Agricultural Organization of the United Nations by Fishing News (Books) Ltd., p. 603–614.

Mosser, J.L., N.S. Fisher, and C.F. Wurster, 1972a, Polychlorinated biphenyls and DDT alter species composition in mixed cultures of algae: *Science,* v. 176, p. 533–535.

Mosser, J.L., N.S. Fisher, and T.C. Teng, 1972b, Polychlorinated biphenyls: Toxicity to certain phytoplankters: *Science,* v. 175, p. 191–192.

United States National Academy of Engineering, 1972, Outer Continental Shelf Resource Development Safety: A Review of Technology and Regulation for the Systematic Minimization of Environmental Intrusion from Petroleum Products, p. 32–134.

Oil Pollution: Persistence and Degradation of Spilled Fuel Oil: 1972 *Science,* v. 176, p. 1120–1122.

Parricelli, J.D., and V.F. Keith, 1974, Tankers and U.S. energy situation—An economic and environmental analysis: *Marine Technol.,* p. 340–364.

Pearson, C.S., 1973, Extracting rent from ocean resources: Discussion of an neglected source: *Ocean Devel. Internatl. Law Jour.,* v. 1, n. 3, p. 221–237.

Portmann, J.E., 1972, Possible dangers of marine pollution as a result of mining operations for metal ores: *In:* Ruivo, M., Ed., *Marine Pollution and Sea Life.* Published by arrangement with the Food and Agricultural Organization of the United Nations by Fishing News (Books) Ltd., p. 343–346.

Pratt, S.C., S.B. Saila, A.G. Gaines Jr., and J.E. Krout, 1973, *Biological Effects of Ocean Disposal of Solid Waste.* University of Rhode Island Sea Grant Publication 73-9, Narragansett, R.I., 53 pp.

Ross, D.A., 1977, *Introduction to Oceanography,* 2nd Ed, Englewood Cliffs, New Jersey, Prentice Hall

Ross, W.M., 1973, *Oil Pollution as an International Problem.* Seattle: Univ. Washington Press.

Ryther, J.H., 1970, Is the world's oxygen supply threatened? *Nature,* v. 227, p. 374–375.

Ryther, J.H., and W.M. Dunstan, 1971, Nitrogen, phosphorus and eutrophication in the coastal marine environment: *Science,* v. 171, p. 1008–1013.

Sanders, H.L., 1977, The West Falmouth spill—Florida, 1969: *Oceanus,* v. 20, n. 4, p. 15–24.

Schachter, O., and D. Serwer, 1970, *Marine Pollution Problems and Remedies.* New York: UNITAR Research Reports, No. 4

Schindler, D.W., 1974, Eutrophication and recovery in experimental lakes: Implications for lake management: *Science,* v. 184, p. 897–898.

Segar, D.A., P. Hatcher, and G.A. Berberian, 1974, *The Chemical and Geochemical Oceanography of the New York Bight Apex Region as it Pertains to the Problem of Sewage Sludge Disposal.* U.S. Department of Commerce, NOAA, Washington, D.C. 67 pp.

Sellers, J.W., 1973, Solid waste disposal and ocean dumping: *Marine Affairs Jour.,* n. 1, p. 52–77.

Stegeman, J.J., 1977, Fate and effects of oil in marine animals: *Oceanus,* v. 20, n. 4, p. 59–66.

Stewart, R.J., 1977, Tankers in U.S. waters: *Oceanus,* v. 20, n. 4, p. 74–85.

Travers, W.B., and P.R. Luney, 1976, Drilling, tankers and oil spills on the Atlantic outer continental shelf: *Science,* v. 194, p. 791–795.

Turekian, K.K., 1974, Heavy metals in estuarine systems: *Oceanus*, v. 18, n. 1, p. 32–33.

*United States Environmental Protection Agency, 1975, *Oil Spills and Spills of Hazardous Substances*. Washington, D.C., 29 pp.

United States National Academy of Sciences, 1971a, *Chlorinated Hydrocarbons in the Marine Environment*. A report prepared by the Panel on Monitoring Persistent Pesticides in the Marine Environment of the Committee on Oceanography, National Academy of Sciences, Washington, D.C., 42 pp.

United States National Academy of Sciences, 1971b, Marine Environmental Quality: Suggested Research Programs for Understanding Man's Effect on the Oceans. Special study held under the auspices of the Ocean Science Committee of the NAS-NRC Ocean Affairs Board, 9–13 August 1971, Washington, D.C., 107 pp.

United States National Academy of Sciences, 1975, *Assessing Potential Ocean Pollutants*. Washington, D.C., 438 pp.

United States National Academy of Sciences, 1975b, *Petroleum in the Marine Environment*. Procedures of Workshop on Inputs, Fates, and the Effects of Petroleum in the Marine Environment, Washington, D.C.

United States National Academy of Sciences, 1975c, *Petroleum in the Marine Environment*. Washington, D.C., 107 pp.

United States National Science Board, 1971, *Environmental Science: Challenge for the Seventies*. Report of the National Science Board, National Science Foundation, Washington, D.C. 50 pp.

Vaughn, J.M., 1974, Human viruses as marine pollutants: *Oceanus*, v. 18, n. 1 p. 24–28.

Vetter, R.C., and W. Robertson IV, 1974, Predicting ocean pollutants: *Oceanus*, v. 18, n. 1, p. 17.

Wastler, T.A., and L.C. Wastler, 1972, Estuarine and coastal pollution in the United States: *In:* Ruivo, M., Ed., *Marine Pollution and Sea Life*. Published by arrangement with the Food and Agricultural Organization of the United Nations by Fishing News (Books) Ltd., p. 40–59.

Weidemann, H., and H. Sendner, 1972, Dilution and dispersion of pollutants by physical processes: *In:* Ruivo, M., Ed., *Marine Pollution and Sea Life*. Published by arrangement with the Food and Agricultural Organization of the United Nations by Fishing News (Books) Ltd., p. 115–121.

Wilson, R.D., P.H. Monaghan, A. Osanik, L.C. Price, and M.A. Rogers, 1974, Natural marine oil seepage: *Science*, v. 184, p. 857–865.

Woodwell, G.M., P.P. Craig, and H.A. Johnson, 1971, DDT in the biosphere: Where does it go? *Science*, v. 174, p. 1101–1107.

Wurster, C.F., 1971, Aldrin and Dieldrin: *Environment*, v. 13, n. 8, p. 33–45.

Zapata, M.C., and R.W. Hann Jr., 1977, Technical and Philosophical Aspects of Ocean Disposal. Texas A & M Sea Grant Publication 78-203, 160 pp.

Chapter 7

Military Uses of The Ocean

Introduction

The oceans have been used for military purposes at least since the times of the early Egyptian Dynasties. Perhaps surprisingly, however, the oceans in a military sense offer some peaceful advantages not found with land-based operations. For example, before radar, two navies could easily pass each other (or even pass through each other) without being aware of each other's presence. With radar and other methods of detection, this is now rarely possible. Ships nowadays may closely tail or "challenge" each other (like a marine version of chicken) with little chance of this leading to any armed confrontation. In most instances the important thing is who controls the seas or, in other words, who maintains the largest fleet in an area with sufficient logistic backup. Nuclear submarines and their detection are probably the most important aspects of the military use of the sea.

A navy has advantages over other types of military forces; for example, it is much easier, both physically and logistically, to move ships than it is to move an army. A navy can travel to ports without having to traverse land, whereas an army's mobility is considerably more limited—it should be mentioned again that the surface of the world is 72% water. An air force has limitations because of fuel needs and the necessity for a friendly landing space (which, of course, could be an aircraft carrier). Another important difference between land and sea operations is that foreign vessels have the right of innocent passage through territorial seas, whereas no such provision exists for armies on land. In the 1958 Convention on the Territorial Sea and Contiguous Zone, innocent passage was defined as not being prejudicial to the peace, good order, or security of the coastal state. Clearly, various interpretations are possible concerning the term "prejudicial to the peace" of the coastal state. The recent series of Law of the Sea negotiations will introduce some restrictions on the movement of navy ships. These are discussed in a later section of this chapter.

Part of the present balance of military power between the United States and the USSR is because both have worldwide ranging missile-carrying nuclear submarines. The large number of these submarines and their awesome nuclear payloads give them destructive "secondary strike" capabilities. Because both sides have them and no dependable method of destroying each other's exists, it can be argued that the submarines prevent either side from initiating a nuclear attack. It is therefore beneficial to maintain this balance rather than change it, which could then lead to a different and accelerated type of development or deployment. In this regard, a restriction or strong regulation on the passage of nuclear submarines through international straits might not, in the long run, be to the benefit of any country. A restriction could come from the Law of the Sea Conference by extension of territorial seas from 3 to 12 naut. mi. If this occurs, over 100 presently international straits will come within the territorial waters of coastal states. In this situation military vessels would, under the present law of the sea, only have the right of innocent passage, and submarines would have to traverse these straits on the surface. The United States has argued for the right of submarines to pass through these straits submerged and for full freedom of overflight by airplanes. When submarines travel on the surface, they are exposed both to conventional hazards, such as collision with surface ships, and to observation.

Military ships are, of course, viewed with special concern by many coastal states. These types of ships can easily be considered as a threat; indeed, the moving of military ships to or near an area is often done exactly for that purpose. A 3- or 12-mile territorial sea is little protection against such a threat. However, to exclude "innocent passage" of military ships through the 200-mile economic resource zone of the world's coastal countries would almost effectively neutralize the advantage of having a surface navy, assuming that such rules were recognized and respected.

At present there is only one major treaty relevant to the military use of the ocean—the Treaty on the Prohibition of the Emplacement of Nuclear Weapons and other Weapons of Mass Destruction on the Seabed and the Ocean Floor and in the Subsoil Thereof. It is commonly called the Sea Bed Treaty and went into effect on 18 March, 1972 after the required number of countries (22) has signed the agreement: the United States, Soviet Union, and Great Britain were among the signatories. The treaty, in spite of its impressive title, is really extremely limited. Basically, it prohibits the placing of nuclear weapons or other instruments of mass destruction on or in the sea floor beyond the outer limit of the territorial sea or contiguous zone (i.e., outside of 12 naut. mi). It does not restrict any actions inside of this zone. The Soviet Union had proposed the outright banning of all military equipment on the sea floor, but this was not acceptable to the United States. The value of permanently mounted nuclear systems on the sea floor is not very significant—at least according to United States military people. It can be said, if one wants to take a pessimistic view, that this treaty bars something of only limited interest that may never be developed. It offers no restrictions on nuclear-equipped submarines, antisubmarine warfare systems, or other underwater military detection systems. On an optimistic point the treaty does have an article encouraging future negotiations concerning the militarization of the ocean. In addition, the treaty itself has value as a step toward a peaceful approach to the military use of the ocean.

In recent years there has been an increased militarization of the ocean. One of the principle reasons for this is that the superpowers have used the oceans as a location for their major weapons systems, because the ocean may provide some invulnerability for these systems. In other words, the countries feel that in the ocean they can maintain their "second strike" capability. In theory the deployment of these weapons systems (mainly on nuclear submarines) could stabilize the arms race and make future arms limitations more possible. Second strike capability can be increased by having more warheads per missile, more missiles per submarine, more range per missile, and deeper firing depths for the missile and by any improvement in the self-defense of the submarine. The technique against nuclear submarine capabilities is called antisubmarine warfare, or simply ASW.

Antisubmarine Warfare

Antisubmarine warfare is a multibillion dollar industry. Technologically, however, it has not kept pace with the increasing sophistication of nuclear submarines. It has been compared to a game of baseball where the outfielders wear blindfolds. The situation is tolerable because it applies to both sides.

It appears that the United States, even before it was a nation, may have been the first country to use a submarine against another country. This happened during the Revolutionary War, when the United States *Turtle* tried but failed to sink the British *Eagle*—the attack took place in New York Harbor. The first sinking of a ship by a submarine occurred in 1864, when the Confederate submarine *Hunley* sank the U.S.S. *Housatonic* in Charleston Harbor. Unfortunately, the submarine was also destroyed by the explosion.

The development of nuclear submarines has led to a whole new area of technology for their detection, monitoring, and possible destruction; however, the advantage still rests with the submarine. Table 7-1 shows some of the

Table 7-1. ASW versus the Submarine[a]

	Submarine				
	Operational Strengths	Weaknesses	Weapons	Weapons Range	ASW Weapons Platform
1776	Virtual undetectability	Range, surface and Submerged speed, surface and submerged endurance, navigation, safety, communication, depth	Mines	Local	Ship
World War I	Virtual undetectability	Submerged speed, submerged endurance, safety, depth	Torpedos Guns	Local Local (surfaced only)	Ship ASW submarine
World War II	Very limited detectability	Submerged speed, submerged endurance, depth	Torpedos Mines Guns	Local Static implantment Local (surfaced only)	Ship ASW submarine Aircraft
Present	Limited detectability	Slight increase in radiated noise and sonar target size compared to World War II	Missiles Torpedos Mines	Local to intercontinental Local Static implantment	

[a]Adapted from Connelly (1976).
[b]Historical—not real time.

Table 7-1 (*continued*).

ASW Platforms and Weapons			ASW Detection, Classification and Localization Systems		
Platform Range	ASW Weapons	Weapons Range	Methods	Effectiveness	
				Short Range	Long Range
Long	Mines	Static	Visual	Poor	Nil
			Intimate contact	Excellent	Nil
Long	Depth charges	Short	Visual	Poor	Nil
	Torpedos	Short	Radio direction	Good	Fair
	Mines	Static	findings		
Long	Torpedos	Short	Acoustical	Poor	Nil
	Mines	Static implantment	(nonelectronic)		
Long	Same as World War I, with Improved Technology		Visual	Poor	Nil
			Radio direction findings	Poor to fair	Poor to fair
Long	Same as World War I, with Improved Technology		Magnetic anomaly	Poor	Nil
			Active sonar	Poor to good	Nil
			Passive sonar	Poor to good	Nil
Long	Bombs (surfaced only)	Short			
	Torpedos	Short			
	Depth charges	Short			
	Mines	Static implantment			
	Same as World War II with Vastly Improved Detail Technology		Active sonar	Fair to excellent	Poor to good
			Passive sonar	Fair to excellent	Poor to good
			Magnetic anomaly	Poor	Nil
			Infrared	Fair to good[b]	Nil to poor

historical changes in the ASW versus submarine contest. For an ASW system to be even barely effective, a basic understanding of sound transmission in the ocean is necessary. This involves knowledge of the daily and seasonal temperature and salinity values of the ocean as well as of the sound characteristics of the ocean bottom.

During World War II the German submarine fleet was destroyed, in large part, because they could not remain submerged and be undetected for long periods of time. With the invention of the snorkel and later nuclear-powered submarines, it became possible for these types of submarines to remain submerged and relatively quiet for longer times. Nuclear submarines are essentially independent of the surface because they do not need oxygen for their engines' propulsion and they and other modern submarines can maintain communications while under water.

Even though some ASW systems are fairly sophisticated, it appears that the modern submarine is relatively safe in the ocean. It can be argued that perhaps ASW technology should not be pushed much further, because there are benefits that arise when both superpowers are unable to detect and destroy each other's submarines. For example, both would then have and be able to maintain a very strong deterrent to nuclear attack, and future arm escalation could be halted or reduced. If one side were to make a substantial ASW breakthrough, it certainly would lead to accelerated action by the other side.

The detection of submarines is a very difficult and complicated procedure that has a low rate of success. Three basic techniques can be used: sonar, magnetic anomalies (because the submarine is composed of steel it will have a magnetic value), and infrared tracking (to detect the heat emitted from the submarine). Sonar is most effective because of the basic characteristics of the submarine—its physical shell will reflect sonar signals and its internal machinery will make detectable sound, as will the cavitation from its propeller when it is moving.

Underwater Sound

The propogation or movement of sound in the ocean is complex, because sound velocity changes with changing temperature, salinity, and pressure (depth). These three components vary vertically in the ocean, horizontally (in the case of temperature and salinity) from area to area, and seasonally (temperature and salinity) because of heating, cooling, and evaporation. Sound velocity will increase 1.3 m/s for each one thousand part increase in salinity, about 4.5 m/s for each degree centigrade increase in temperature, and 1.7 m/s for each 100-m increase in depth. To calculate sound velocity (which must be known for ASW work) the values of these variables must be determined. A series of general correction factors has been developed for

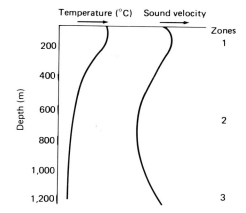

Figure 7-1. A typical sound velocity profile in the ocean. The arrows indicate increasing values. The three different zones are discussed in the text.

most areas of the ocean, but for accurate work, sound velocity measurements must be made at the time of the military operation.

Because of the vertical changes in sound velocity the ocean can be divided into three basic zones (see Fig. 7-1):

1. A surface zone of well-mixed water, where sound velocity increases mainly because of the pressure or depth effect. This zone is generally about 100–150 m in thickness.

2. An intermediate zone, where the sound velocity will decrease because of the temperature decrease. This zone can extend to a depth of 1500 m.

3. A deep zone, where sound velocity increases because of increasing pressure (the temperature and salinity are essentially constant). This zone generally extends to the ocean floor.

This distribution of sound changes, combined with the fact that sound waves—like ocean waves (Fig. 8-4)—can be refracted, produces two militarily important sound areas, called shadow zones and sound channels.

A shadow zone is an area where it is hard for sound to penetrate. This zone generally occurs in the upper parts of the ocean, where an increasing sound velocity gradient overlies a decreasing sound velocity gradient and the sound source is in the increasing sound velocity (or upper) layer (see Fig. 7-2). In this fairly common situation, sound is refracted upward in the increasing sound area and downward in the decreasing sound area (sound is refracted toward the area of lower sound velocity), producing the so-called shadow zone. As can be seen from Fig. 7-3 it can be very difficult to detect a submarine situated in a shadow zone.

A sound channel occurs if there is an area (in the vertical sense) where the sound velocity values reach a minimum. Sound in this minimum value area

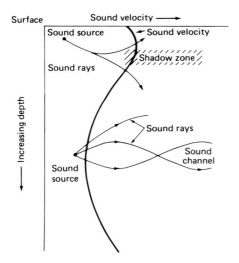

Figure 7-2. A shadow zone and a sound channel.

will be refracted either upward or downward to the area of lower velocity or, in other words, back into the minimum value area, forming the sound channel (see Fig. 7-2). The energy loss of sound with distance tends to be considerably lower in this sound channel, and sound transmission of literally thousands of miles is possible within the channel. A sound velocity minimum is commonly found at a depth of about 1500 m and is called the SOFAR channel. Sound transmissions of over 15 000 miles (about 28 000 km) have been made through this channel. This SOFAR channel may be used by ships in distress, for an explosive charge detonated in the channel would be detected by many listening stations and could be used to determine the approximate position of the ship. It is this principle that may have been used to locate the United States nuclear submarine *Scorpian* when it sank.

In general as sound travels through the ocean, it loses or decreases in energy because of three factors: spreading, absorption, and scattering.

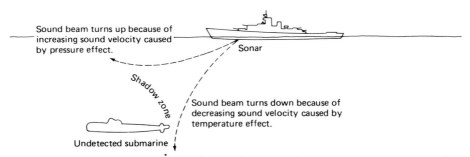

Figure 7-3. How a shadow zone, formed because of sound refraction, can provide an area of relative safety for a submarine.

Sound loss from spreading is simply proportional to the square of the distance that the sound travels. Water, itself, will absorb sound and convert it to heat; this absorption is proportional to the square of the sound frequency—the higher the frequency, the greater the absorption. Finally, sound can be scattered by organisms and particles, including gas, in the water and by the ocean bottom.

ASW Systems

Sonar or acoustic techniques to search for submarines can be used by helicopters, surface ships, airplanes, submarines, and buoys, or even by systems fixed on the sea floor. Ocean-bottom systems have been operational for many years, and basically they can be of two types: either passive or active. A passive system uses a series of hydrophones placed on the bottom that continuously listen for sounds from submarines or other vessels. Hydrophones are devices that respond to underwater sound and generate an electrical signal. These devices can be emplaced almost anywhere on the ocean bottom or can be extended through the water from buoys. Because their positions are known it is relatively simple to use a series of these emplacements to accurately locate and follow a particular ship. An active surveillance system consists of transmitters that actually produce sound signals, which in turn are bounced off or reflected from other ships or submarines in the area; the returning energy is used to pinpoint or detect the vessel.

As early as 1967 a passive surveillance system was known to have been installed on the east coast of the United States at a water depth of about 600 ft (about 182 m)—a similar system was probably in place along the Pacific. This system, known as the Ceasar Program, undoubtedly has been improved considerably and probably additional systems are in place in other areas of the ocean. The new systems, of which little is known, are named Colossus, Barrier, and Bronco.

In a military sense, fixed-bottom installations, besides being outlawed in certain areas under the Sea Bed Treaty, are at a considerable disadvantage when compared to mobile systems that can be used by submarines, surface ships, or aircraft. For example, a series of sonobuoys (sound detecting devices) can be deployed by surface ships or airplanes and, by using computer techniques, it may be possible to discriminate, from all the other signals in the ocean, the presence of a submarine. The advantage of using one submarine to hunt for another is that submarines are considerably more quiet than surface ships, thus producing less interfering noise. However, the quieter the enemy submarine, the more difficult it is to detect.

A common system is for ships routinely to tow hydrophones. It has been reported (*Booda* 1975, p. 17) that one of these hydrophone systems on a United States ship has detected a submarine at a distance of 6000 miles (over 11 000 km) from the towing ship. Any listening device will constantly receive a large variety of signals, including those from fish, mammals, wind (which form waves), surface ships, submarine volcanic activity, and an occasional submarine. Shipboard computers and knowledge of the characteristics of this background noise (everything but what you are looking for) are obviously necessary for successful ASW work.

The actual operating depth of nuclear submarines is classified. One prototype vessel, a deep-diving U.S. Navy submersible called *Dolphin,* has the capability to go 10 000 ft (3048 m). This nonnuclear submarine can carry a crew of over 20 and is used to test deep-ocean systems and weapons. A nuclear powered deep-diving submersible, the *NR-1,* has a seven-man crew and has been used for oceanographic research.

Often there are newspaper or other reports of new antisubmarine weapons developed by either the United States or the USSR. One such United States device is called "the Captor," which is a "sleeping" type of torpedo that lies on the sea floor. It is activated by the sound of an approaching enemy submarine and then silently homes in on and destroys the enemy ship. If it misses on the first try, it supposedly will turn and try again. The torpedo is programmed not to respond to a friendly submarine and can be positioned on the sea floor by submarine, surface ships, or even airplanes. It appears to be a very formidable weapon to exclude submarines from a certain area, but probably only under conditions of war.

There is a constant technologic battle between the navies of the United States and the Soviet Union to obtain even the slightest ASW advantage. A recent newspaper report (Boston Globe, 19 Dec. 1975) noted that the Soviets were using a new resin on the outsides of their submarines that was absorbing sonar signals rather than reflecting them. The net effect is to make the detection of these submarines more difficult. However, United States sources indicated that new sonar techniques were already being developed to overcome this problem. United States authorities also claim to have a complex and elaborate underwater network system of listening devices that can track Soviet submarines in most of the important areas of the ocean. This method of location, as elaborate as it is supposed to be, is still far from being adequate for actual destruction. New missile devices, such as the cruise missile, which can fly near the ground below enemy radar, following the terrain by TV and hitting its target with amazing accuracy, are thought to represent the ultimate deterrent.

Both countries will go to rather extreme efforts to learn more about each other's capabilities. One of the more amazing of these efforts is that of the *Glomar Explorer,* built for the Summa Corporation, owned by Howard Hughes. For several years it was thought that the Summa Corporation was going to be in the manganese nodule business. They had built a 618-ft (188-m) long, 115.5-ft (35.2-m) wide, 36 000-ton experimental ship called the *Glomar Explorer* (Fig. 7-4). The cost of the vessel and the rest of the system was on the order of half a billion dollars. Numerous press releases

Figure 7-4. The *Glomar Explorer*, a vessel thought to be an experimental deep-sea manganese nodule mining ship but later found to be part of an operation to raise a sunken Russian submarine. (Photograph courtesy of Summa Corporation)

(although always with some element of mystery) told how the ship would bring nodules up from the bottom by air suction and store them in a 324-ft (98.7-m) barge submerged below the *Glomar Explorer*. In early 1975, however, it was learned that the Hughes' venture was really a cover for a CIA attempt to raise physically a sunken Russian submarine from the Pacific. It is still not known how successful the recovery was. The operation showed several things: One was that there was American technology available to recover large objects from any portion of the sea floor. It also indicated to the rest of the world, and especially to less developed countries, that what might appear to be a peaceful or perhaps scientific operation in the deep sea could have military implications. This point was not missed by the delegates meeting at the 1975 Law of the Sea Conference, which, ironically, was being held when the Hughes story broke. Ultimately, the *Glomar Explorer* may still be used in a manganese nodule mining operation, as it has recently been leased to a mining consortium for tests in the Pacific (see Chap. 5, p. 143).

As previously noted, there are some good military reasons to halt or control future ASW developments. There is a precedent for such an action in the agreement between the United States and the Soviet Union concerning their landbased missile systems—each has agreed not to build large ABM (antiballistic missile) systems for defense against their systems. If such a

Table 7-2. Strengths of Some Navies

Country	Aircraft Carriers	Cruisers	Destroyers	Ballistic Missile Submarines	Cruise Missile Submarines	Fleet and Patrol Submarines	Missile Boats	Others[b]
Canada	0	0	4	0	0	3	0	98
China	0	0	3	1	0	57	100	1768
Egypt	0	0	5	0	0	12	12	86
France	2	2	20	4	0	19	1	354
Israel	0	0	0	0	0	2	18	62
United Kingdom	3	12	1	4	0	29	0	522
United States	21	27	107	41	1	77	0	551
USSR	2	35	108	72	68	248	135	1670

[a]Data from *Jane's Fighting Ships* (1975–1976).
[b]This number includes a wide variety of naval ships, including patrol boats, mine sweepers and layers, landing craft, maintenance ships, supply ships, oilers, and research ships.

similar approach were applied to the ocean it might reduce the military use of the ocean. Two possible suggestions are:

1. Make large areas of the ocean only available to either the United States or the Soviet Union—conceivably that area off the other country's coast.

2. Prohibit large sonar arrays in the ocean.

The effect of these two suggestions could be to give each country considerable confidence in their own marine nuclear deterrent—if they do not have it already.

Relative Naval Strengths of the United States and the USSR

The Soviet Union is clearly one of the world's great maritime powers. It excells in several areas, including size of navy (Table 7-2), size of merchant fleet (see Table 4-1), oceanographic research ships, and fishing boats. Comparisons are sometimes difficult because of differing criteria; I have presented two tables, one from the authoritative *Jane's Fighting Ships* (Table 7-2) and a shorter one prepared by the U.S. Library of Congress in 1976 and published in several news (Newsweek, 1 March 1976) magazines (Table 7-3).

Estimates of the relative military mights of the United States and the Soviet Union are common, the purposes of which range from political to budgetary urgencies. Nevertheless, the Soviet Navy in terms of ships alone does outnumber the United States fleet and perhaps of more importance is developing a worldwide network of bases. The United States has chosen to build large attack carriers, which although easy to track and perhaps destroy, allows the U.S. Navy a considerable air power advantage. It is also generally thought that the U.S. Navy is considerably better trained than its Soviet counterpart.

Table 7-3. United States–Soviet Union Naval Comparison (1976)

	United States	USSR
Major combat ships	182	226
Aircraft carriers	14	1
Missile submarines	41	73
Attack submarines	73	253

In 1978, Admiral Holloway III, Chief of Naval Operations, indicated that there had been a reversal of the pattern of recent years, with the U.S. Navy now getting an increasing number of modern war ships while the Soviet Navy is starting to become obsolete. The reason for this change in trend, according to Holloway, is that the U.S. Navy retired many of its vessels at the end of World War II, which reduced the size of the fleet, but its recent shipbuilding program is putting new ships into service. The Soviet Union, in contrast, did not have a big fleet after World War II and began building many new ones at that time; these ships are now becoming obsolete.

The United States is presently converting several of its *Polaris*-type submarines into *Poseiden*-type submarines that are equipped with MIRV (multiple independently targetable reentry vehicles) missiles. Each will contain 16 missiles, with each missile containing 10 independently targetable nuclear warheads. It is believed that these submarines are essentially invulnerable within the present state of the ASW art. Nevertheless, a new system of weapons—the *Trident*—is being developed. These submarines will have an Undersea Longrange Missile System (ULMS) on longer range, deeper diving submarines. Each *Trident* will carry 24 missiles that have a range of 4500 nautical miles (over 8300 km). These submarines will probably each cost more than $1 billion—30 such submarines are planned. This system is thought to be sufficient until the year 2000. It is believed that the Soviet Union is also developing advanced missile systems.

The United States and the Soviet Union, in their 1972 Strategic Arms Limitation Treaty (SALT) agreement, agreed to limit the amount of submarines and missiles that each country can have. The understanding is to hold the number of submarine-launched missiles and modern missile submarines to the numbers operational and under construction at the signing of the agreement. The limits are: for the United States, 710 submarine-launched missiles and 44 modern missile submarines; for the Soviet Union, 950 submarine-launched missiles and 62 modern missile submarines. This agreement does not limit additional modernization and replacement of present missile systems or submarines; in addition, there are no limitations on the number of warheads carried by the missiles.

In this regard the *Trident* missile submarine system is not really restricted by the SALT agreement, because fewer than 44 are planned and the total missiles they could carry can be below the maximum number allowed. Again, however, and more important, there are no restrictions on the number of multiple warheads that the missiles can carry.

The U.S. Navy has generally been the dominant force in the Pacific Ocean, whereas while the Atlantic Ocean has been generally "controlled" by the United States and other NATO countries. In contrast, the Indian Ocean has essentially been devoid, until recent years, of any dominant naval force. This situation has changed recently, with the British abandoning some of their Indian Ocean bases and the Soviet Navy expanding in the area. The United States is expanding its Indian Ocean facilities, in particular with the development of a base on the island of Diego Garcia, about 1000 miles (1800 km) south of India. It should be noted that with so many supertankers coming and going from the Persian Gulf area, that control of shipping in this area is especially critical because of the world's increased dependence on oil.

For many years, the Soviet Union has been handicapped by only having ports either in cold water or adjacent to narrow straits (see next section). Recent political arrangements, however, have yielded them several warmwater ports, such as in Cuba, Ethiopia, Angola, and one temporarily in Egypt.

Military Use of the Ocean and the Law of the Sea

The last few years have seen several "limited wars" (Korea, Vietnam, and the Arab–Israeli conflicts to name a few) in which naval action mainly has been used to support land operations; submarine warfare has not been involved. In these types of conflict the neutrality and territorial claims of nearby countries were generally strongly maintained; however, it should be noted that military operations could be performed outside the presently recognized territorial sea (generally 3 or 12 naut. mi.) and easily could reach the adjacent land. An interesting exception occurred prior to the 1967 Arab–Israeli conflict, when Egypt imposed a blockade over the Strait of Tiran and the Gulf of Aqaba. The issue was extremely complex, but the objective was to prevent shipping from reaching the Israeli port of Elat. The blockade was broken by a United States flag-flying tanker. Subsequently Israel got control over the strait and until recently has resisted its return. An extension of territorial seas or restriction of military ships from a 200-mile economic resource zone could have a considerable effect on such "limited wars," assuming that such restrictions would be honored. Some military actions almost have occurred on the high seas; the best example of this was the 1962 Cuban missile confrontation, when the United States threatened to stop the search Soviet ships on the high seas.

"Showing the flag" is a common ploy used by some navies. If done with the host country's approval, an extension of territorial seas is irrelevant. However, it probably would serve little purpose if the warships had to stay 200 naut. mi. off the coast. Navy ships have been used to help confirm the freedom of the seas in controversial areas. An example was Great Britain's use of its ships in the mid-1970s to challenge Iceland's extension of its territorial seas to 50 miles in what was called the "Cod War."

Some narrow straits have special regulations concerning military vessels (the importance of straits in general has been mentioned earlier). The Dardanelles, the Sea of Marmara, and the Bosporus, which together connect the Black Sea to the Mediterranean, have their passage rules governed by a 1936 convention. This convention permits the passage of warships from Black Sea states (with the exception of aircraft carriers and submarines) through the area without hinderance. Light warships from other countries can pass subject to restrictions on the length of stay, number of ship already in the area, and size of the vessel. The passageway is essentially controlled by Turkey and, in wartime, if Turkey is neutral, the straits are closed to all warships—a point that could have considerable impact on the Soviet Union, which maintains a considerable fleet in the Black Sea.

The Soviet Union has shown special concern in certain areas, such as the North Sea, about the presence of oil rigs. Apparently, it is feared that the rigs could contain equipment for monitoring the movement of Soviet submarines. In early 1976, several Soviet ships were "chased away" from some North Sea oil rigs. One of the most important pieces of water for the Soviet Union and its allies is the exit to the Baltic Sea. Over 14 000 Communist ships pass between a narrow 8-mile (about 15-km) wide passage off Denmark, called the Great Belt. In addition, an estimated 80% of Soviet and its allies' ship-building and ship-repair facilities are located on the Baltic. These straits, besides being narrow, are also shallow and easily can be mined, a point well understood by NATO and Warsaw Pact countries. The Soviets generally keep several vessels on patrol near the area.

The influence of a nation's claim to a portion of the ocean will be determined by the strength of that nation and by the type of warfare that is occurring. Obviously, in a total nuclear war little regard would be given to territorial seas, economic resource zones, etc. Small countries realize, however, that in time of peace an increase in territorial claims will, as a net result, decrease or restrict the power of large navies. It is very clear that future naval strategy will be influenced by the events coming from the Law of the Sea Conference. Indeed, the military groups of most major countries have had considerable input into developing their states' negotiating position at the conference.

Several articles in the Law of the Sea negotiations specifically relate to military ships. For example, in the territorial sea submarines must navigate on the surface and show their flags. Military ships in the territorial sea must not do anything prejudicial to the peace, good order, or security of the coastal state (so-called innocent passage) or exercise or practice any type of military activity. Military ships will also have innocent passage rights through archipelagic waters. No regulations have been proposed for the high sea, nor is it probable that any could even be enforced.

References

*Booda, L.L., 1975, Top level defense guidance in ASW grows: *Sea Technol.,* November, 1975, p. 16–18.

Booda, L.L., 1976, Undersea warfare improved effectiveness awaits major top level decisions: *Sea Technol.,* November, pp. 11–13, 40–41.

Brown, E.D., 1969, The legal regime of inner space—Military aspects: *Current Legal Problems,* v. 22, p. 181–204.

Brown, E.D., 1971, *Arms Control in Hydrospace: Legal Aspects.* Oceans Series 301, Woodrow Wilson International Center for Scholars, Washington, D.C., 131 p.

Caldwell, T.F., 1975, Strategic consequences of antisubmarine warfare: *Sea Technol.,* November, p. 11–13.

*Connelly, W., 1976, Sonobuoy–Aircraft combination nullifies submarine's advantages: *Sea Technol.,* February, p. 16–19.

Cruise missiles: Air Force, Navy weapon poses new arms issues: 1975, *Science,* v. 187, p. 416–418.

Gallup, E.L., 1973, Sovereignty of the seas and the effect upon naval strategy: *Marine Affairs Jour. Univ. Rhode Island,* v. 1, p. 1–8.

Garwin, R.L., 1972, Antisubmarine warfare and national security: *Sci. Amer.,* v. 227, No. 1, p. 14–25.

Glomar Explorer: CIA's salvage ship a giant leap in ocean engineering: 1976, *Science,* v. 192, p. 1313–1315.

Hirdman, S., 1975, Prospects for arms control in the ocean: *In:* Borgese, E.M., and Krieger, D., Eds., *The Tides of Change; Peace, Pollution and Potential of the Ocean.* New York: Mason/Charter, p. 80–99.

Hirdman, S., 1971, Weapons in the deep sea: *Environment,* v. 13, p. 28–42.

Hull, E.W.S., 1976, The International Law of the Sea: A Case for a Customary Approach. Occasional Paper No. 30, Law of the Sea Institute, University of Rhode Island, Kingston, R.I.; p. 5–7.

Janes Fighting Ships (1975–1976), J. Moore ed. New York, Franklin Watts Inc.

Janis, M.W., and D.C.F. Daniel, 1974, The USSR: Ocean Use and Ocean Law. Occasional Paper No. 21, *Law of the Sea Institute, University of Rhode Island,* Kingston, R.I.

Rubin, A.P., Sunken Soviet submarines and Central Intelligence laws of property and the agency: *Amer. Jour. Internatl. Law,* v. 69, p. 855–858.

Sulikowski, T., 1975, Soviet ocean policy: Ocean Devl. Internatl. Law Jour., v. 3, p. 69–73.

Trident in trouble: New missile may resemble Poseidon, after all: 1976, *Science,* v. 191, p. 50–51.

*References indicated with an asterisk are cited in the text; others are of a general nature.

Chapter 8

The Coastal Zone

Introduction

As far as people are concerned the coastal zone is probably the most important part of the ocean as about two-thirds of the world's population lives near the coast. The coastal zone sits at the interface between the two major environments of the earth—land and ocean. It can be defined as that part of the ocean affected by the land and that part of the land affected by the ocean. The two environments meet at the shoreline, a constantly changing position. In the seaward direction, certainly part, if not all, of the continental shelf should be included within the coastal zone, whereas in the landward direction estuaries, marshes, sea cliffs, and other similar environments are included. Actually there is no universally acceptable definition of the coastal zone. Some have suggested that the entire 200-mile exclusive economic zone (see Chapt. 3) be considered as part of the coastal zone.

Definitions and the terminology for the numerous features within a typical coastal zone are often confused. I have adapted the terminology of Inman and Nordstrom (1971) as shown in Fig. 8-1. The coastal zone includes, besides the continental shelf, more inland features, such as estuaries, lagoons, bays, deltas, and coastal plains. A shoreline and shore zone can also be defined within the coastal zone. The shore zone covers that area where water and land come in direct contact and includes the beach and surf zone. The shoreline marks the point where land and water meet; the nearshore region is that area seaward of the shoreline. The shoreline is a very dynamic area that, on a worldwide basis, has a length of over 400 000 km.

The areal extent of the coastal zone can also be considerable. Inland

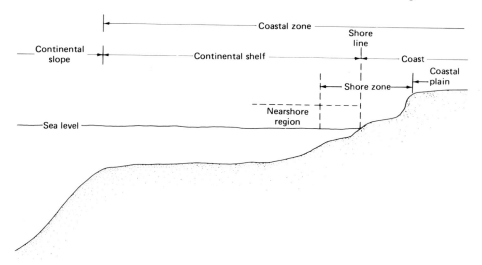

Figure 8-1.   General features of the coastal zone. (Adapted from Inman and Nordstrom 1971)

waters, such as semienclosed bays, estuaries, and lagoons, exceed 100 000 km² for the United States alone. The area seaward of the shoreline, the continental shelf, covers a worldwide area of about 50 million square kilometers.

It is hard to underestimate the importance of the coastal region to the general aspects of the oceans.   For example, an extremely large number of marine organisms either originate or spend a major portion of their lives in coastal estuaries or marshes, and even more inhabit the continental shelf. As indicated in Chapter 5, over 90% of our fishing products come from areas that are included within the coastal zone.   The coastal zone can easily be considered as the bottom step of the ocean's ecologic ladder.   Rivers enter the ocean through the coastal zone, carrying both beneficial products, such as nutrients, and damaging ones, such as pollutants and waste products from coastal industries and municipal dumping.   These materials will eventually be distributed and mixed into the ocean by such coastal processes as waves, currents, and tides.

Ownership and Definition of Coastal Regions

The ownership of coastal areas was a legal issue as far back as the time of the Magna Carta.   It was established then, as common law, that lands over which the tides flowed were owned by the throne and were held by it for the

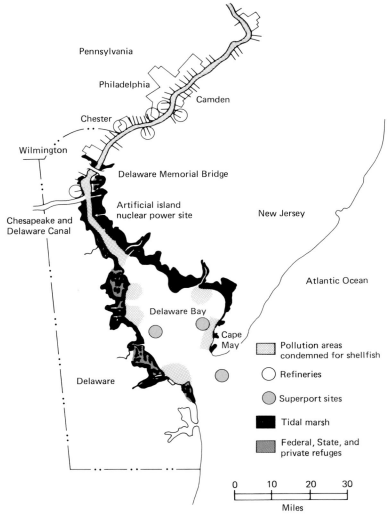

Figure 8-2. The Delaware estuary and its many and sometimes conflicting uses. (Adapted from *Mosaic*, 1973, a publication of the National Science Foundation)

use of the public. This concept of ownership of tidal lands was incorporated into American law when the United States was formed. Eventually, however, the Supreme Court modified this doctrine and said that upon the formation of the United States the 13 states received the ownership of these areas that had been previously held by the British throne.

The exact location of the shoreline is very important for determining property boundaries as well as for ascertaining the beginning of various offshore boundaries (i.e., the territorial sea and economic resource zone). The shore and the shoreline, now as throughout the geologic past, will always be an area of change. In some instances the ocean is winning by erosion; at other times land builds seaward by deposition. At certain times the sea level may rise or fall dramatically, because of either global glacial periods or local tectonic movements. There is little, if anything, that humans can do to prevent these changes. Nevertheless, the exact position of the shoreline is important. Some countries have adopted the seventeenth century English rule concerning marine property lines, which says that the "ordinary" high-water mark (i.e., high tide) is the boundary between the public and private domain. It should be appreciated that the tidal range can vary over a short distance and that closely spaced measurements may be necessary to determine accurate tidal heights. The height of the sea surface, besides being affected by tides, can be influenced by winds, currents, bottom topography, atmospheric conditions, local obstructions, etc. Generally, five tidal data are used: mean higher high water (MHHW), mean high water (MHW), mean tide level (MTL), mean low water (MLW), and mean lower low water (MLLW). Often it is the mean high water or MHW that is used as the boundary between sovereign state land, subject to private ownership, and the tide lands, owned by the state (Houlder 1976). The MLW generally marks the lower boundary of the tide lands and is often the baseline from which boundaries in a seaward direction are measured. These rules, however, can vary considerably. For example, on the west coast of the United States the MLLW line is used, whereas on the east coast the MHW is used.

In the United States authority over the near-shore parts of the coastal zone is generally divided between different Federal, state, and local agencies. Many of these agencies administer overlapping and commonly conflicting uses of the coastal area. Local governments generally have control over land use and waste disposal, whereas state agencies can control water, pollution control, highways, and ownership of state lands. These different patterns of ownership and fragmented control can lead to complexities and incorrect decisions concerning the coastal zone (Fig. 8-2). Before we consider some of these problems, a brief discussion of oceanographic processes in the coastal zone is appropriate.

Oceanographic Processes Relative to the Coastal Zone

Probably the most important marine process relative to the coastal zone is that resulting from the motion of water, in particular, waves produced by wind. The exact process whereby energy is transferred from wind into waves is not well understood; however, the general characteristics and effects of waves are. Wind waves, similar in shape to a sine wave, have a height, length, period (the time it takes for two waves crests to pass a given point), and calculable velocity.   In deep water (i.e., well away from the shore or areas where waves are breaking) it is the wave form that is actually moving forward, not the water itself, which essentially just moves up and down. This can be visualized by holding a piece of paper between one's two hands and moving one hand up and down, creating a wave that moves across the paper. This is substantially the motion that the water is making in deep water.

The basic characteristics of waves, such as their height and length, result from three variables: the velocity of the wind, the duration of time that the wind blows, and the fetch or the distance over which the wind blows.   It is because of the last variable that waves in small lakes and ponds can never really become very large, because the fetch is always small in these areas compared to the open ocean.   In contrast, the wind over the ocean can blow over very long distances and generate extremely large waves.   Waves generated in the Antarctic region (where the wind can blow around the whole circumference of the earth) can travel across the entire Pacific, Atlantic, or Indian Oceans, breaking perhaps on Hawaii or elsewhere.

Waves can be in one of three main classes, called sea, swell, and surf (Fig. 8-3).   Sea waves are those in the actual area of a storm and are very irregular, with a confused pattern.   As the waves leave the storm area, however, they travel with a velocity that is proportional to the period or the length of the waves.   Waves with similar periods or similar lengths therefore will travel together.   Longer period waves, all things being equal, will travel faster than shorter period waves.   When waves of similar character are together, this is called swell.   Swell is what many visualize as a typical image of the ocean—the broad, long-period waves.   All waves will eventually come to a shallow area and break, and when this happens they are called surf.   The force of these waves can be considerable; it has been estimated that a 3-m high wave, per unit length, contains essentially as much power as a large automobile running at full throttle.

When waves break there is a basic change in their physical characteristics that causes a definite forward motion of water, resulting in considerable energy moving toward the beach.   As the waves get near shallow water, several things happen that have important effects on the near-shore region. Generally when waves approach a coast, each individual one has a uniform wave height along its crest.   This is because of a flow of energy along the crest of a wave at right angles to the direction at which the waves approach the coast.   The effect, called diffraction, can be extremely important if there is an obstruction or breakwater that interferes with the wave form.   When this happens there is a flow of energy back along the wave crest that tries to reestablish the original wave height.   This may cause movement of the bottom sediment in the same direction and possibly the silting-in of the area behind the breakwater.

Figure 8-3, a–c. The different types of wind-generated waves. **a** Sea waves in the generating or storm area. **b** Swell; note that the wave pattern is relatively uniform. **c** Swell or breaking waves. (Ross 1977)

a

b

c

Probably the most important thing that happens as waves approach shallow water is refraction. Refraction caused by bottom topography will cause waves to change their direction of approach to the coast. The main result of refraction is that the waves turn toward the shallower water so that the wave crests tend to parallel the depth contours. If there is an offshore submerged bar, projecting point, or cliff, therefore, the wave energy converges toward it, whereas it diverges over a submarine canyon or deeper area (Fig. 8-4). The principal effect of waves approaching a coast and breaking is to straighten out the coastline, either by erosion or by deposition, or both. Building of breakwaters, piers, and jetties will interfere with this effect and disturb the equilibrium of the dynamic processes in the near-shore environment. This is of course, sometimes the objective of building such structures; often it is not.

After waves break, the forward movement of the water is toward the

Figure 8-4. Aerial photograph showing wave refraction. The waves are traveling from the lower left to the upper right. Note the convergence, caused by refraction, of the waves toward the land. (Photograph courtesy of U.S. Air Force, Cambridge Research Laboratories)

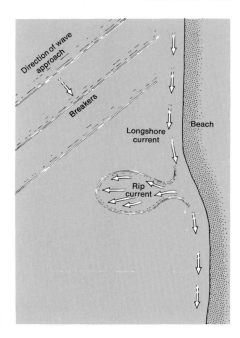

Figure 8-5. Longshore currents and rip currents (See Fig. 8-6) in the near-shore region.

beach. The beach, however, is an essentially impermeable barrier. If the waves approach the beach at any angle, the discharge of water from the breaking waves is directed along or parallel to the beach, eventually forming a longshore current (see Fig. 8-5). This longshore current can transport sediments parallel to the beach, but eventually the water in the current builds up so that it must move seaward. When the water moves seaward, a rip current (Fig. 8-6) is produced; its position is dependent upon the bottom topography and the height and period of the waves. Rip currents can carry sediments seaward, but they can again be moved landward with the next sequence of approaching waves. Thus, waves approaching a coast will essentially keep the bottom sediments in somewhat of a transient state: eroding in one place, depositing in another, but eventually smoothing out the coastline—unless humans interfere.

Longshore currents often will cause a net along-shore movement of the sediment and can eventually straighten out a shoreline. This process can take thousands of years and often causes considerable local erosion. In general, longshore currents and storms can erode beach sediments and can damage private property. Owners often construct jetties or groins that cause the deposit of the sediment carried by the longshore current (see Fig. 8-7 and 8-14). Downstream of the barrier erosion occurs, a process that often leads to the construction of another groin or jetty, and so on.

The circulation in the near-shore zone results in an exchange of the water

Figure 8-6.   A series of rip currents, which can be distinguished by light sediment-laden water moving seaward beyond the breaking waves. (Photograph courtesy of D.L. Inman, Scripps Institution of Oceanography)

Figure 8-7.   Photograph showing how a series of jetties have trapped the sand moving along the beach (from upper right to lower left). Note how the area to the lower left of the largest jetty is relatively narrow and has little sand. This is because the sand that normally would have been carried to this area has been intercepted by the jetty.

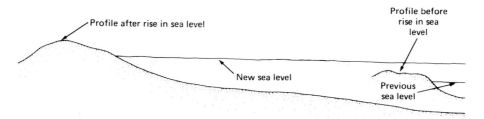

Figure 8-8a.   Changes in a beach because of a long-term rise of sea level.

in this region with the more offshore areas. In this manner nutrients, pollutants, etc., and other material carried by the rivers are eventually distributed
offshore.

Coastal Erosion

A 1973 U.S. Corps of Engineers study showed that excluding the Alaskan
coast, about 42% of the United States shoreline is eroding. The erosion is
most critical in the North Atlantic area, where population is densest and
where up to 85% of the shoreline is privately owned. Part of the coastal
erosion problem around the world is related to the general slow rise in sea
level because of glacial melting (Fig. 8-8a) and subsidence of coastal areas in
some localities. The important factors affecting coastal erosion are, accord-

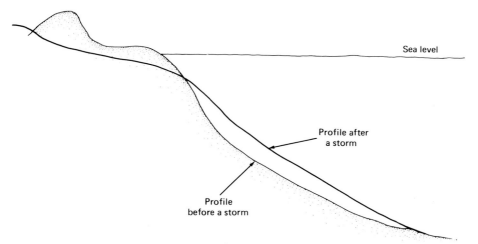

Figure 8-8b.   Changes in a beach from local storms. (Adapted from Pilkey and
others 1975)

ing to Inman and Brush (1973): degree of exposure to waves and currents; supply of sediment and runoff to the coast; shape of the coast and adjacent continental shelf; tidal range and its intensity; and coastal climate. Storms or hurricanes can have an especially destructive effect on beaches (Fig. 8-8b). Human activities and construction, however, probably have had a more profound recent influence on coastal areas.

Erosion is generally especially pronounced on beaches and coastal cliffs. Changes will tend to be less where the coast is rocky or where the beach sediments are mainly gravel. Especially dramatic effects can occur along gently sloping sandy beaches. The fine grain size of the sediments allows them to be more easily moved by waves, currents, and especially storms. Beaches are relatively unstable landforms and usually will show a distinct seasonal change, generally being eroded in winter by short storm waves and built up in the summer by relatively calm seas, which move sediment landward (Fig. 8-9). This effect results from the fact that the relatively long-period waves of summer pick up sand in shallow water and carry it onto the beach. The sand is then carried back seaward by the backrush of water from the wave running down the beach; however, because the backrush is smaller than the incoming waves the sand is not carried as far seaward as its original position. The following wave again causes a net forward motion toward the beach. In wintertime, however, waves tend to be of shorter period, which keeps the sand more in suspension rather than letting it settle as in the case of the longer period wave. Therefore, much of the sand carried off the beach by the backrush does not settle until it is carried into deeper water, beyond the

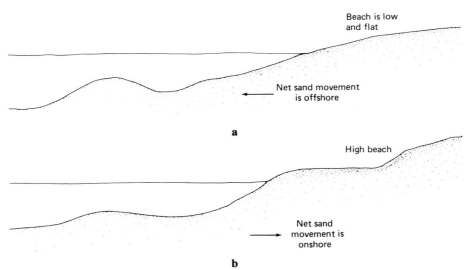

Figure 8-9, a and b.   Typical seasonal changes along a beach. **a** Winter. **b** Summer.

action of subsequent waves. This will generally cause a temporary loss of the beach sand.

Beaches also are changed with tides, storms, and the action of humans. The alternating erosion and deposition of beaches is a natural phenomenon which, over long periods of time, and if not interfered with, will produce minimal changes. It is just this interference which will cause increased erosional problems. Basically there are three ways that humans can affect coastal and shoreline areas: land recovery by dredging and filling, damming rivers, etc.; construction of jetties or other coastal structures; and the development or destruction of coastal dune areas. Coastal dunes, if left alone, will act as a preventive barrier to inland erosion; but when they are removed or breached erosion can occur. Dunes are especially sensitive to abuse and can be easily damaged by vehicular traffic, paths, or roads. Any of these activities breaks down the plant structure which has anchored the dune, and once a gap exists in the dune pattern storms can penetrate and erode them.

Estuaries and Marshes

Several other coastal environments are rather unique and deserve special care. These include estuaries and wetland regions, such as salt marshes and mangrove swamps (Fig. 8-10). An estuary can be roughly defined as that part of a river where tidal and river currents interact and mix. They generally are areas of quiet water with intermediate salinity. Sedimentation rates are high because of the material carried in by the rivers. Most estuaries have resulted from the recent rise of sea level (about 130–140 m in the last 15 000 years) that has "drowned" the rivers and created the estuary. Environmentally and esthetically, estuaries are very appealing areas and seven of the world's 10 largest cities are sited in such regions.

Unfortunately, the same properties that make estuaries unique biologically also have led to their industrial development (see Fig. 8-2). Estuaries are generally very rich in nutrients in large part because of the rivers that enter them. Some waste products can be nutrient-rich and actually beneficial to the environment. The nutrient supply can support a large population of phytoplankton and provide a major food supply for organisms higher up in the food chain, such as zooplankton, fish, and shellfish. In most instances, estuaries are the major habitats and area of highest productivity of organisms along the coastal region. Many fish or other creatures will spend a portion of their life histories in coastal estuaries. It has been suggested that more than 50% of the commercially valuable fish spend part or all of their time in estuaries.

Somewhat similar to estuaries are lagoons, which are broad, shallow areas partially restricted from the ocean by offshore barrier beaches. If the open-

Figure 8-10. Air view of a salt marsh on Cape Cod, Massachusetts. This entire marsh is drained by a single major channel. Although no major freshwater stream enters this marsh, generally the main channel is a river or estuary. (Photograph courtesy of Dr. Ivan Valiela)

ing or access to the ocean gets closed, lagoons can eventually become freshwater lakes.

Many estuaries or coastal areas are bordered by a narrow vegetation or wetland area containing either salt marshes or mangrove swamps. Mangroves are generally found only between 30°N and 30°S, or in the equatorial region, whereas marshes occur in the more temperate climates (Fig. 8-11). This distribution, it is speculated, is controlled by the apparent inability of mangrove seedlings to survive freezing temperatures.

Salt marshes are generally located in intertidal areas, along the banks of tidal rivers, or behind barrier beaches. They are an extremely productive area and support a large marine population, including fish, birds, shellfish and plants. Marshes also form a protective barrier against storms and high seas for the land behind them. The organic production of marshes, increased by nutrients supplied by rivers, can be as high as 5–10 tons per year of organic matter per acre compared to 1 ton per year per acre for a wheat field or less than 0.5 ton per year for the open ocean or for a desert. This large production of organic matter results in food that can be consumed by the myriad of organisms that inhabit the area, either for part or for all of their lives.

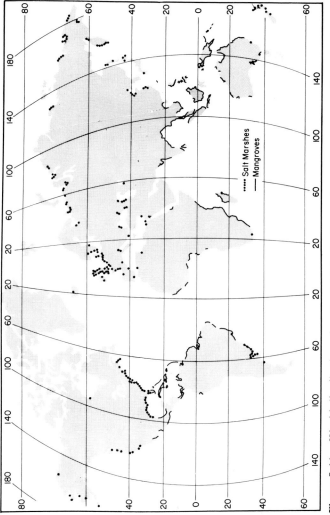

Figure 8-11. World distribution of well-established salt marshes and mangroves. Because the data are mainly from the literature, all such areas are not represented. (Valiela and Vince, 1976)

Figure 8-12. Industrial development in a salt marsh area. In the foreground is a waste disposal area that is slowly covering the marsh. (Photograph courtesy of Dr. Ivan Valiela)

Because wetlands are in prime development areas, there are considerable pressures to develop them. In the State of Connecticut, for example, over 50% of the original marshland has been destroyed and of the 14 000 or so acres remaining in 1969, about 200 or so acres are being filled in each year (Fig. 8-12). As the cities grow, in some localities, the marshes are slowly built over and developed (Fig. 8-13).

Marshes commonly contain many food products, such as clams, oysters, scallops, and fish, that are important to the local economy. They can also have other important values; for example, marshes can remove some pollutants from the water, in particular, nitrogen and some metals. Experiments are being conducted to see the effects of adding sewage to coastal marshes. A value additional to removing the pollutants is that nutrients from the sewage increase the productivity of the marsh. The nutrients can be used by the plants and lead to their increased growth and thus to an increased source of food to other organisms in the food chain.

Estuaries and lagoons, like many other parts of the coastal area, are really only transient or temporary features. With time, they will be destroyed either by the cutback of the coastline caused by marine erosion or by the build-

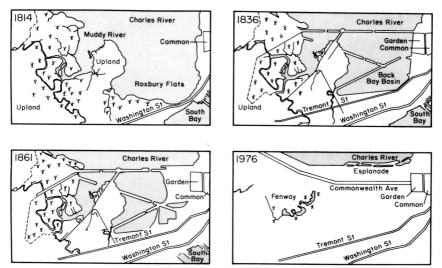

Figure 8-13. A series of charts showing the development and filling in of the Back Bay area of Boston from 1814 to 1976. In 1814 the main part of Boston was situated in the northeast. By 1836 filling was done to support railroad tracks over the Roxbury flats. Today, almost all the marsh and shore areas are gone. Bunches of reeds indicate salt marshes. Shaded areas are water and unshaded areas are uplands. (Valiela and Vince 1976)

up of marshes. By their growth, marshes will tend to close the seaward parts of the estuary. The rivers entering the estuary also tend to fill it up, as could sediments supplied from offshore areas. If the estuary is an important navigational area, therefore, it may need to be kept open by constant dredging.

Human Activity in the Coastal Zone

Many of the problems in the near-shore region come from the building of coastal structures that interfere with circulation of water and sediment and thus cause increased deposition in some locations and erosion elsewhere. These coastal structures, described below, especially can interfere with longshore transportation and prevent replenishment of beaches (Figs. 8-7 and 8-14). This results in erosion on the downstream side of the structure and accretion on the upstream side. The effect can be compounded by a series of structures—a result that generally occurs because the owner of the downstream property is "forced" into taking some sort of action to prevent

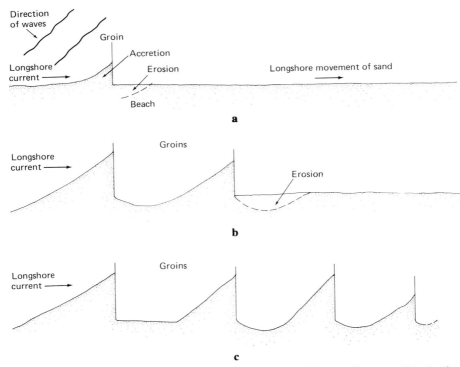

Figure 8-14, a–c. Plan view of a sequence of events that can happen with the installation of groins into a longshore current system. Once one groin is added (**a**), others will become needed (**b** and **c**). (Adapted from Inman and Brush 1973)

the now increased erosion of his or her beach. It perhaps is hard to visualize the amount of sand that moves along a beach, but many ocean beaches can have as much as 0.5–1 million cubic meters of sand transported across them in an average year.

The most common types of coastal structures or mechanisms to prevent erosion are:

Groins: Structures built perpendicular to the beach into the water that interrupt the longshore movement of water and sand.

Seawalls: rigid structures built parallel to beaches to withstand and reflect oncoming waves (Fig. 8-15).

Breakwaters: offshore, massive structures situated parallel to but off the shore that absorb the breaking of the wave before it reaches the shore. They generally produce a quiet zone behind them that serves as an accumulation place for sand (and so cause erosion elsewhere).

Figure 8-15. A series of houses built dangerously close to a beach; note the rocks and the stone wall placed so as to reduce erosion. The pile of material in the right foreground is seaweed deposited by a previous high tide.

Revetments: layers or blankets of strong, nonerodable material placed on the shore to prevent erosion.

Artificial fill: instead of trying to prevent erosion, fill material is added (often yearly) to replenish the beach.

Often a combination of techniques may be used.

Some new and innovative structures have recently been built to diminish and dampen the effects of waves and other forms of energy that cause coastal erosion. These include floating breakwaters that can adjust their direction with changing currents or winds. Such devices, if big enough, can provide shelter for even large vessels, although none such has yet been built. Perhaps more realistic are groups of tires tied together that can act as a simple, but cheap, breakwater (Fig. 8-16). Another idea came from a group of scientists at Scripps Institution of Oceanography headed by Professor John Isaacs. The idea, simply stated, was that ocean waves would lose much of their energy if they were to meet rows of spherical balls attached to the sea floor but floating close to the surface (Fig. 8-17). Preliminary tests show that the system is effective enough to remove up to 50% of the wave energy. The floats behave like upside down pendulums that have a back and forth movement faster than that of the incoming waves. The floats therefore move in the direction opposite to the waves and dissipate much of the energy or, as Isaacs says, "the system beats the waves to death." Such a system costs only a small fraction of conventional breakwaters and offers considerably more flexibility. For example, it can be moved.

Any attempt to prevent shoreline protection or restoration should be

Figure 8-16.   Groups of scrap tires tied together to form a cheap, but efficient, floating breakwater. (Photograph courtesy of the University of Rhode Island)

Figure 8-17.   A tethered float breakwater designed to dissipate the energy of wind- or boat-generated waves in near-shore water. (Official U.S. Navy Photograph)

made with considerable thought and, if possible, research. The conditions initially observed might not represent the average or even typical situation.

Use of the Coastal Zone

Most of the problems relative to the use of the coastal zone would be solvable if they were independent of each other; unfortunately, however, they are not. Many aspects of one use of the coastal zone will have a negative or positive effect on another use. Consider, for example, two sets of conflicting possible usages:

1. Developing offshore or near-shore deep-draft oil terminals versus protection and conservation of the shore zone for recreation and the preservation of its natural resources.
2. Development of waterfront homes (Fig. 8-15) or larger developments that will destroy, damage, or change the beachfront area versus preserving these areas both for their esthetic value and for their "breakwater" ability for more inland areas.

The number of possible uses of the coastal zone is almost endless (Table 8-1). In some instances, multiple usage can be accomplished, but this requires knowledgeable management, skill, and compromises between the different interests.

To get a better view of the demands on the coastal zone, I list below some important considerations about this region in the United States; however, many points also relate to other coastal areas in the world:

—Seventy percent of the United States population lives within a day's drive of a coastal region.

—According to the U.S. Department of the Interior, only about 6.4% of the United States shoreline that can be used for recreation is under public ownership; the rest is privately owned. The contiguous states have about 60 000 miles of shoreline, only about one-third of which has some recreational potential.

—Concerning marine recreation, in 1972 a total of $3.9 billion in retail sales was generated for coastal zone use and there were 9.2 million boats, 80% of which were in the 30 coastal states. By the year 2000 the number of boat owners is predicted to more than double.

—Seven of the largest cities in the United States are on the coast and over 33% of that country's population lives in coastal counties, which only contain about 15% of the total United States land area. Actually, the five largest cities in the world (Tokyo, New York, London, Shanghai, and Osaka) are found in coastal areas.

Table 8-1. Coastal Zone Problems for a Local Area of Long Island, New York as Identified by the Regional Marine Resources Council of the Nassau–Suffolk Regional Planning Board[a,b]

---

 1. Reduction of commercial shellfish production
 2. Depletion of sport and commercial fin fisheries
 3. Control of insects and related pests
 4. Disposal of solid wastes
 5. Destruction of wetlands
 6. Development of marine industries
 7. Stabilization and protection of the coastal shore
 8. Dredging and spoil disposal
 9. Eelgrass
10. Disposal of domestic wastes
11. Boat pollution
12. Oil spill pollution
13. Limited shoreline recreation facilities
14. Duck waste pollution
15. Saltwater intrusion into freshwater supplies
16. Thermal pollution
17. Preservation of sites of natural or historic value

---

[a]From Heilcoff (1976).
[b]Note that many of these problems are interrelated.

—Shoreline property has been continuously increasing in value, usually at more than 10% a year, and in some areas coastal shorefront property costs are excessive except for only the wealthy. Clearly, the price of this land will not decrease in the future, because such land can only get scarcer.

—About 40% of United States manufacturing takes place in coastal counties; coastal facilities handle about 350 million tons of foreign trade and 620 million tons of domestic cargo (Ducsik 1974).

—The ownership of the United States shoreline (excluding Alaska) is as follows: 70% private, 12% state or local government, 11% Federal, and about 3% uncertain.

—Many large energy facilities, oil refineries, and an increasing number of nuclear facilities are situated on or very near the coast. Many of these facilities, especially the nuclear plants, require large daily supplies of water.

—All ships engaged in marine transportation spend a significant portion of their time in harbors (generally estuaries) within the coastal zone.

—Over 100 million United States people participate yearly in ocean-related activities, spending close to $15 billion dollars in the process.

—Close to 50% of the United States coastline is suffering severe erosion.

—Coastal development has already destroyed or built over a large percentage (probably greater than 20% and less than 50%) of our coastal habitat areas.

—Daily waste disposal from municipal and industrial activity is about 30 billion gallons. Dredging spoils amount to about 100 million tons per year.

—A very large percentage of commercial fish spend a major portion of their life cycles in the coastal zone. Shellfish, such as clams, oysters, and mussels, spend their entire life cycles in the more restricted waters of estuaries, bays, marshes, etc.

—Coastal waters receive the major portion of waste material from land, including sewage; radioactive and thermal discharge; and dredging, mining, and construction spoils.

There are authorities who feel that the destruction of the coastal zone in many regions of the world has proceeded so far that it can never be returned to its original state. Certainly in some areas it will be impossible to restore the coastal zone to the original pristine character it had before human arrival and development of this area (Fig. 8-18). A more realistic goal would be to

Figure 8-18. Aerial view of Port Newark and Elizabeth Port Authority Marine Terminal, an extremely developed coastal area. (Photograph courtesy of the Port Authority of New York and New Jersey)

keep pollution and destruction to a minimum and to make choices the long-term effects of which will be beneficial to as many people as possible. This will mean a nuclear power plant in one place and an undeveloped coastal park in another, rather than one of either of these everywhere.

One of the major problems of the coastal zone is the discharge of sewage and other types of effluent into it. Among the important aspects of an offshore effluent system are what happens to the material once it reaches the water, how far away the effluent still can be detected and therefore be considered a pollutant, how much effluent can be safely discharged into an area, and whether it has any effect on public health and the esthetic aspects of the near-shore waters. There are several ways to attempt to answer these questions, including modeling, putting dye in the water and noting its movement, and measuring the characteristics of bottom sediments in the water itself.

An example of the size of such a problem can be seen in the area off New York City (called the New York Bight), which received an average of 4.6 million metric tons of solid material per year during the late 1960s. About 76% of this material was dredge waste, 12% came from construction and demolition rubble, 7% was solids and waste chemicals, and 4% was from sewage sludge. The material was discharged over an area of about 160 km^2 in the city's harbor and over 50 km^2 offshore on the continental shelf.

The problem for the coastal zone is clearly one of wise management, and a major part of that is jurisdictional. Local governments , of course, will see only the local view, whereas the National Government (in theory) should be considering a broader picture. Decisions concerning the coastal zone are naturally influenced by the potential and present resources of the area. The value of ocean resources, most of which are in or closely related to the coastal zone, is immense. According to a recent study, by the year 2000 this economic potential could exceed $33 billion for the United States (Table 8-2).

There are many reasons that decisions concerning the coastal zone have been delayed or incorrect. Among the more obvious are: the conflicting jurisdiction and lack of coordination between governmental agencies; lack of a clear plan for the development of the coastal zone (see next paragraph); lack of knowledge and data concerning the resources and marine interactions of the coastal zone; lack of funds to manage the coastal zone; and conflict between economic uses and ecologic uses.

A leader in the concern for the coastal zone has been the United States, which in 1972 passed the National Coastal Zone Management Act (P.L. 92-583). This act encourages individual coastal States to exert a strong role in all matters effecting the coastal zone. An Office of Coastal Zone Management was established to make Federal funds available to a state if it developed a comprehensive plan for the use of its coastal zone. As part of this plan the state must make baseline studies of its coastal resources, define problem areas, and develop the appropriate political mechanisms for a management program. When the management program of the state is completed

Table 8-2.   Estimated and Projected Primary Economic Value of Selected Ocean Resources to the United States by Type of Activity 1972/1973–2000, in Terms of Gross Ocean-Related Outputs (in billions of 1973 dollars)[a]

Activity	1972	1973	1985	2000
Mineral resources				
Petroleum		2.40	9.60	10.50
Natural gas		0.80	5.80	8.30
Manganese nodules			0.13	0.28
Sulfur		0.04	0.04	0.04
Fresh water		0.01	0.02	0.04
Construction materials		0.01	0.01	0.03
Magnesium		0.14	0.21	0.31
Other			0.01	0.02
Total		3.40	15.82	19.52
Living resources				
Food fish	0.74		0.95–1.58	1.37–4.01
Industrial fish	0.05		0.05–0.08	0.05–0.14
Botanical resources	[b]		[b]	[b]
Total	0.79		1.00–1.66	1.42–4.15
Nonextractive uses				
Energy			0.58–0.81	3.78–6.03
Recreation	0.70–0.97		1.12–1.50	1.64–2.53
Transportation	2.57		4.40–6.21	6.88–11.41
Communication	0.13		0.26–0.36	0.44–0.85
Receptable for waste	[c]		[c]	[c]
Total	3.40–3.67		6.36–8.88	12.74–20.82
Grand total	7.59–7.86		23.18–23.36	33.68–44.49

[a]Source: United States Senate Committee on Commerce (1974).
[b]Insignificant.
[c]Potentially significant but unmeasurable.

it must be approved by the Secretary of Commerce before the state receives funding.

Under the program, states first receive planning grants to develop some sort of management program. Once these are approved, the states can then receive further grants, called administrative grants. Several requirements are necessary for the beginning of a coastal zone management program: First, the local governments must define the area of their coastlines for management purposes. Generally this extends out to 3 miles off their coasts and includes intertidal areas, beaches, dunes, and wetlands as well as inshore areas to that extent to which the shoreline would affect coastal waters. Within these regions the state has to designate areas of particular concern for one reason or another. These can be oil and gas considerations, environmental problems, recreation potentials, etc. Second, a state has to designate and control those uses in a coastal zone that can affect the coastal waters. As the states start to develop their management programs, they must collect information concerning the resources of the coastal zone and then develop a plan based on that information and show how the coastal zone should be used and developed. The final step is to develop a management technique to implement the plane.

In this process the states must develop specific programs, preparing maps, illustrations, etc., concerning the coastal zone, and define the areas that are relevant to the management program. They must develop objectives and policies to guide public and private use of the lands and waters in the coastal zone itself. There must be an inventory of the areas involved, and guidelines for use of the areas and how they will be used in the future.

Public hearings must be held as the program is developed and the program must be approved by the governor of the state and a state agency must administer the grants for the development of the program. As these mechanisms are developed there will be considerable input from different user groups within a state and, in particular, within a coastal zone. This will involve adjustments and changes to the program as it is developed.

In 1976 the President of the United States signed a new 10-year program (Coastal Energy Impact Program) to assist states and regions affected by offshore energy activity; it contained amendments to the 1972 Coastal Zone Management Act. The amendments provided increased financial assistance to states and local governments toward alleviating the impact of offshore oil and gas activity on the coastal zone. In all, $800,000,000 was authorized for such federal loans, which could be repaid by grants at a later time. An additional $400,000,000 was authorized for grants to be given relative to exploration, development, and production of oil in the area and to pay for unavoidable losses of recreational or ecologic facilities. The new amendments also guaranteed that offshore activities were consistent with state and local coastal zone standards and that the public would continue to have access to coastal and beach areas. The Coastal Energy Impact Program began in 1977 and nearly $5 million was awarded to states.

In the United States the interaction of the coastal zone and offshore oil and gas development can have considerable impact. It is not just the drilling that presents the problems, but the fact that personnel are needed for drilling, and that ship facilities, support industries, and staging areas are necessary along the coast for offshore operations. Large amounts of supplies and materials will be moved and generally helicopter facilities must be available. Refineries, pipelines, and storage tanks may also be built in the near-shore area. All this development will put considerable strain on existing harbor facilities. Likewise, if people come to work in these areas, they will need homes, schools, and other facilities, which will make yet an additional impact on what probably has been a relatively small community. The local areas can rightly argue that the oil companies, to get offshore concessions, pay immense amounts of money directly to the Federal government with little, if any, coming directly back to the state or local government. Indeed, any oil drilling 3 miles or more off a coast falls into the Federal regime and local governments will not directly receive any of the royalties associated with the drilling. Obviously, in some instances, the increase in personnel in the industries associated with drilling can be an advantage for the local community. However, in many instances it can strain the existing situation.

Two other important United States programs for the coastal zone have been developed; one is the Marine Protection Research and Sanctuaries Act of 1972. A provision of this act permits the Federal government to designate specific areas of the coastal zone from the high-tide line out to the edge of the continental shelf for research protection and recreational purposes. The first such national marine sanctuary was the site of the sinking of the Civil War vessel U.S.S. *Monitor* off Cape Hatteras. The second area was a marine park in the Florida Keys, and by 1977 over five areas covering 25 000 acres had been designated as sanctuaries.

The second important act is the Fishery Conservation and Management Act, passed in April of 1976 and implemented in March of 1977. This act has given the United States a good start in conserving and managing the fish resources off its coast. A total of eight Fishery Management Councils have been formed and they are developing management plans for their geographic areas (see Chap. 5 for more details).

References

Biggs, R.B., 1975, Offshore industrial-port islands: *Oceanus*, v. 19, p. 56–66.

*Brahtz, J.F.P., (ed.), 1972, *Coastal Zone Management: Multiple Use with Conservation*, New York: John Wiley and Sons, 352 pp.

Cruickshank, M.J., and H.D. Hess, 1975, Marine sand and gravel mining: *Oceanus*, v. 19, p. 32–44.

Dahle, E. Jr., 1973, *The Continental Shelf Lands of the United States: Mineral Resources and the Laws Affecting their Development Exploitation and Investment Potential.* University of North Carolina Sea Grant Publ., 73-11, Chapel Hill, N.C., p. 27–28.

*Ducsik, D.W., 1974, *Shoreline for the Public: A Handbook of Social, Economic, and Legal Considerations Regarding Public Recreational Use of the Nation's Coastal Shoreline.* Cambridge, Mass.: MIT Press, p. 32–46.

Ducsik, D.W., 1975, *Teaching Coastal Zone Management: An Introductory Subject.* Rept. No. Massachusetts Institute of Technology Sea Grant Program 75-1, 144 pp.

Economic Factors in the Development of a Coastal Zone, 1970. A report prepared for the National Council on Marine Resources and Engineering Development, Massachusetts Institute of Technology, Cambridge, Mass., 140 pp.

*References indicated with an asterisk are cited in the text; others are of a general nature.

Englander, E., J. Feldman, and M. Hershman, 1977, Coastal zone problems: A basis for evaluation: University of Washington Sea Grant Publ. 77-8, from *Coastal Zone Management Jour.*, v. 3, p. 217–236.

Flory, J.F., 1975, Oil ports on the continental shelf: *Oceanus,* v. 19, p. 45–55.

Geological Society of America, 1976, Impact of Barrier-Island Development—Geologic Problems and Practical Solutions. Report of the Committee on Environment and Public Policy for the Council of the Geological Society of America, 8 pp.

Hayden, B., and R. Dolan, 1975, *Classification of the Coastal Environments of the World.* Final report prepared for U.S. Office of Naval Research Geography Programs, Washington, D.C., 159 pp.

*Heilcoff, J., 1976, *Politics of Shore Erosion: Westhampton Beach.* Ann Arbor: Ann Arbor Science Press, 173 pp.

Hollings, E.F., 1975, National Ocean Policy: Priorities for the Future: *Oceanus,* v. 19, p. 8–19.

*Houlder, R.H., 1976, Establishing marine boundaries—A new Federal effort: *Sea Technol,* December, p. 33–35.

Inman, D.L., and R.A. Bagnold, 1963, The sea: Ideas and observations 3: *In:* Hill, M.N., Ed., *The Earth Beneath the Sea.* New York: Interscience, p. 529–536.

*Inman, D.L., and B.M. Brush, 1973, The coastal challenge: *Science,* v. 181, p. 20–31.

*Inman, D.L., and C.E. Nordstrom, 1971, On the tectonic and morphologic classification of coasts: *Jour. Geol.,* v. 79, p. 1–21.

Krueger, R.B., 1971, Elements of oceans and coastal zone management: *Undersea Technology* July, p. 37–40.

Library of Congress Congressional Research Service, 1976, *Effects of Offshore Oil and Natural Gas Development on the Coastal Zone.* A study prepared for the *Ad Hoc* Select Committee on Outer Continental Shelf, U.S. House of Representatives. 396 pp.

Massachusetts Office of Coastal Zone Management, 1976, *Living by the Sea: A Massachusetts Citizens' Handbook for Coastal Zone Management Planning.* Boston, 44 pp.

*Pilkey, O.H. Jr., O.H. Pilkey Sr., and R. Turner, 1975, *How to Live with an Island.* North Carolina Department of Natural and Economic Resources, Raleigh, N.C., 119 pp.

Ross, David A., 1977, *Introduction to Oceanography:* 2nd ed., Englewood Cliffs, N.J.. Prentice-Hall, Inc., 438 pp.

Schoenbaum, T.J., 1972, Public rights and coastal zone management: *North Carolina Law Rev.,* v. 51, 44 pp.

Teal, J., and M. Teal, 1969, *Life and Death of the Salt Marsh.* Boston: Little, Brown, 278 pp.

U.S. Army Corps of Engineers, 1973, *National Shoreline Study:* House Document 93-121, 93rd Congress, 1st Session, v. II and III.

United States Senate Committee on Commerce, 1974, *The Economic Value of the Ocean Resources of the United States.* National Ocean Policy Study, 93rd Congress, 2rd Session.

*Valiela, I., and S. Vince, 1976, Green borders of the sea: *Oceanus:* v. 19, p. 10–17.

Chapter 9

Innovative Uses of the Ocean

Instead of attempting to write an entire book on innovative uses of the ocean I have chosen to highlight some of the more probable future possibilities and avoid the impractical and improbable. In particular, I emphasize possibilities for using the ocean as a source of energy, the interaction of the ocean and climate, fresh water from the ocean, the ocean as an ultimate resting place for nuclear wastes, and imaginative uses of, for example, satellites and the organisms in the ocean. It goes without saying that other major innovative uses of the ocean are likely to be developed that have been completely unanticipated in the writing of the chapter.

Energy from the Sea

Anybody who has seen the sea, either from the shore or from a ship, cannot help being impressed by its energy potential. With the oil crisis of recent years and the awareness of the world's shortage of conventional energy, there has been considerable attention directed toward the ocean as a new source of power. Many new ideas have been suggested, and although some will never become operational, the enthusiasm of their supporters has been contagious. Undoubtedly, some of these devices could help alleviate the world's need for energy in the coming decades.

Energy from the ocean can come from several major sources. These include: waves, tides, ocean currents, thermal differences, salinity differences, and marine biomass. The oceans, which cover 72% of the earth, receive the major portion of the solar energy reaching our planet (Tables 9-1 and 9-2). This energy, because more of it reaches the equatorial region than the polar region, produces an atmospheric circulation pattern that in turn causes an oceanic circulation (see Figs. 2-12 and 2-13). Atmospheric winds also cause waves and some currents.

Table 9-1. Estimates of Power Levels in Natural Processes of the Planet Earth[a]

Available Sources of Power	Total Power (W)
Direct solar power	
Where sun hits atmosphere	10^{17}
At earth's surface	10^{16}
Photosynthesis (Stores sunlight in the form of chemical energy in fats, proteins, and carbohydrates, all of which are combustible.)	
Marine plants	10^{14}
Arable lands, forests	10^{13}
Bioconversion of waste materials	
Plant residues and manure (Can be converted by bacteria to gaseous fuels—hydrogen and methane—by storing them in airless containers at proper temperatures.)	10^{12}
Garbage, sewage, and dumps (Can be converted by the same process.)	10^{12}
Ocean thermal power	
Solar heat absorbed by ocean water (Can possibly be put to use by exploiting temperature differences between surface and depths, producing power to drive turbine.)	10^{13}
Steady surface-wind power, such as that from trade winds	10^{12}
Variable surface-wind power (in middle latitudes where winds are unsteady)	10^{12}
Hydroelectric power (from harnessing the kinetic energy of moving waters)	
Power in rainfall (conceivably could be harnessed, but the world's total rainfall, even if the rain dropping on the oceans is included, would satisfy only 10% of the world's power demand.)	10^{12}
Flow of rivers (harnessable by traditional hydroelectric plants)	10^{11}
National evaporative exchanges between large bodies of water (the Mediterranean Sea and Red Sea are examples: evaporation is greater in them than in the ocean at large; therefore, there is a continual flow into them from the oceans to replace evaporated water. This flow can be harnessed just as in a millrace.)	10^{9}

Table 9-1.   Estimates of Power Levels in Natural Processes of the Planet Earth[a] (*continued*)

Damming of evaporative sinks. (By damming ocean openings to Red Sea and Mediterranean Sea, letting these seas evaporate until a drop of 100 m or more occurs, and then letting the ocean flow in, turning mill wheels, additional power might be obtained. It is not very practicable to build these dams, however.)	10^{11}
Tidal flow (Particularly, this may be done at such places as the Bay of Fundy, where flow can be harnessed.)	10^{9}
Power of the great ocean currents, such as the Gulf Stream and Kuroshio Current (Theoretically these can be harnessed the way rivers are, with some sort of "water wheel.")	10^{8}
Ocean surface waves at the coastline (the power of waves is available at a potential total yield of 10^{6} W/km of coastline.)	10^{10}
Geothermal power (this could be harnessed particularly at the "ring of fire" around the Pacific Ocean basin, so called because this is where tectonic plates merge and volcanoes erupt; the same happens along midocean ridges.)	10^{10}

Present Power Demands	Total Power (W)
Worldwide power demand for all needs of civilization	10^{13}
Human metabolism (total power in terms of food needed to sustain present population level of 4 billion)	10^{11}

[a]Information from von Arx (personal communication) and von Arx (1974b & c)

Table 9-2.   Some Energy Units

1 foot-pound	=	1.356 joules
1 calorie	=	4.184 joules
1 British thermal unit (Btu)	=	1055 joules
1 Btu	=	252 calories
1 watt-hour (W-h)	=	3600 joules
1 Kilowatt-hour (kWh)	=	3413 Btu
1 Kilowatt (kW)	=	56.92 Btu/min
1 megawatt (MW)	=	1 million W
1 barrel of oil	=	5 800 000 Btu
1 cubic foot of gas	=	1031 Btu
1 ton lignite coal	=	20–40 million Btu

Energy from Waves

The strength of waves is well known to anybody who lives near the shore. There are numerous stories and much folklore about waves moving boulders, piers, or seawalls weighing hundreds or even thousands of tons over large distances or of such structures' being broken by waves during storms. The maximum heights that sea waves can attain is not known, but estimates of as high as 112 ft (about 34 m) have been made. For the use of energy from waves, however, it is not the storm situation that is important but the normal or average state of the sea that should be considered.

The amount of energy in even small waves is considerable. For example, a 3-m high wave will transmit energy at a rate of 100 kW/m of its crest (Inman and Brush 1973). Such power has been compared to that of a solid wall of automobiles running at full throttle.

According to Richards (1976) there are three principal methods for extracting energy from waves:

1. Techniques that use the vertical rise and fall of the crests and troughs of successive waves to drive an air- or water-powered turbine
2. Techniques that use the rolling motion of waves to move vanes or cams to turn turbines
3. Techniques that converge waves into channels and concentrate their energy

The first method, using wave-activated turbines, has been used for several years to produce small amounts of energy, such as in buoys to provide energy for lights or noise-making devices. Larger systems, some several hundred feet in diameter, have been proposed that would have the waves compress air and drive a turbine. In spite of their size, such systems probably would not generate major amounts of power. Some countries, especially Britain and Japan, are more optimistic about wave power. Both countries have committed several millions of dollars toward research and development efforts. In Japan a floating 500-ton experimental device has been producing 125 kW of power—it uses the up and down motion of waves to trap air that in turn drives turbines, producing electricity.

Lockheed recently has developed and patented a wave energy machine called "Dam Atoll". It is a dome-shaped device about 80m in diameter that sits just below the water surface. Waves, when they reach the Dam Atoll, will enter an opening at the top and move through a central vortex turning a turbine and thus producing electrical power. Since the device uses wave energy the water behind it will have reduced waves creating a beneficial effect to coastal regions. No further details are available at this time.

Isaacs and his colleagues (1976) have devised a wave power pump (Fig. 9-1). The pump consists of a vertical riser with a flapper valve and buoyant float at the surface. It is loosely attached to the bottom and can move direct-

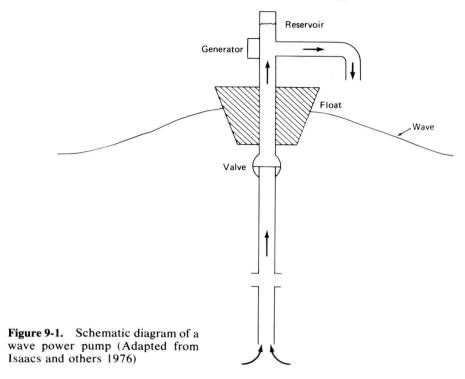

Figure 9-1. Schematic diagram of a wave power pump (Adapted from Isaacs and others 1976)

ly with the motion of the waves. When it is in operation the valve is closed for about half of the wave cycle, allowing the water in the column to move upward with the float. When the float starts to descend with the passage of the wave trough, inertial forces move the water inside the column even higher. As subsequent waves pass, the water continues to rise even higher, eventually being high enough to generate power. A recent test using a 300-ft (about 91-m) long pump magnified the pressure head of the waves over 20 times (Isaacs and others 1976). A series of these pumps could provide a modest supply of energy. Advantages of the system are that it has few moving parts and no fuel costs. These devices could be especially appealing near offshore drilling sites or similar features where a power source is needed.

The second method of generating power from waves, that by using the rolling motion of waves, has not been as well developed. One technique could be a series of vanes that would be moved by wave motion and eventually turn a turbine.

Among the appealing aspects of using waves as a source of energy is that these devices can be used almost anywhere and are nonpolluting. However, they can cause ecologic and environmental problems if they interfere with the breaking of waves, littoral drift, or the movement of sand in the coastal zone (see pages 263–265).

Energy from Tides

People have dreamed of using tidal power since ancient times, and tidal mills have been in operation for centuries. Some harbors built before the time of Christ were designed so that tidal forces would flush them and keep them clear of sediments. Modern tidal energy plants, in spite of the enthusiasm of their supporters, have not been a very effective source of energy. Two plants are presently in operation, one on the Rance River in France (Fig. 9-2) and another in the Soviet Union off the Ura River, but neither is a major power source. One problem with tidal systems is that the tides do not flow continuously but change directions several times a day and vary in strength over a 2-week period (spring and neap tides). This problem can be partially overcome if the water is stored, for example, in a reservoir (Fig. 9-2) at high tide and used to turn turbines either on a continuous basis or when the energy is most needed.

The Bay of Fundy and the Passamaquoddy area of Maine in North America have been considered for several decades as potential areas for tidal energy sites. With each increase in the cost of energy, plans are brought out and rediscussed. Tidal ranges can be as high as 70 ft (about 21 m), and it has been estimated that a 180-MW plant could be built there for

Figure 9-2. The Rance power plant is a 750-m-long structure that has a reservoir of 184 million cubic meters of water. Tides in this region, which can have an amplitude up to 13.5 m, drive turbines (below the surface) during both incoming (rising tide) and outgoing (falling tide) conditions. Energy production can be about 240 MW. (Photograph courtesy of French Engineering Bureau)

about $400 million. This plant, which would produce less than 10% of the energy of a nearby controversial nuclear plant (Seabrook), probably would not be competitive with other sources of energy. Positive points about tidal plants in general are that they will have a longer life than other more conventional energy plans and that tides are free (although if less than 5 m in height they probably are impractical for use). A tidal plant also does not require any major technologic breakthroughs. However, the ecology of the area may be affected by such a plant because it can interfere with or reduce the tidal regime within the estuary.

Energy from Ocean Currents

Ocean currents, such as the Gulf Stream or the Kuroshio Current, off Japan, may also be a future source of energy. Such devices as underwater windmills could be put into a current and by their movement turn or drive a turbine. Other more imaginative devices include a series of parachutes attached to a continuous cable that opens and catches the current but closes when coming back against it. Either of these devices would be difficult to build and maintain in a swift current. Nevertheless, the kinetic energy in a major current is impressive. The Gulf Stream, for example, off Florida carries over 30 million cubic meters of water per second at speeds of almost 3 knots.

Other possible systems include damming up large bodies of water in areas having high evaporation rates. The evaporation will reduce water height on one side of the dam, whereas the water on the other side, if connected to the open ocean, will remain at its original elevation. The result is a hydrostatic head or elevation difference that could be used for power generation. Two possible examples for such an operation are across the Straits of Gibraltar and at the southern end of the Red Sea at the Straits of Bab el Mandab. Obviously, such devices, if only because they would interrupt important ocean commerce, will probably never be built.

Energy from Thermal Differences

The incoming solar radiation from the sun is converted into thermal energy in the ocean and warms the surface waters so that they have a higher temperature than the waters at depth. These two water layers are generally separated by a zone of rapid temperature change, called the thermocline (Fig. 2-16). This temperature difference or thermal gradient can be a potentially valuable energy source.

One of the more interesting and controversial innovative uses of the ocean is that of ocean thermal energy conversion (OTEC). Basically, this is a system that uses the temperature differences between the surface and deeper parts of the ocean to generate power. In doing so, it essentially needs no fuel and could be a major source of energy for the future. The ocean is the main energy source for the system. Warm water is drawn in from the surface and cold water is taken in from depth. When ammonia, or whatever working fluid is used, comes in contact with the warm sea water it evaporates into a gas that can be used to drive the turbine of an electrical generator. This technique is similar to that used in conventional electricity generation. The colder sea water, pumped up from depth, is then used to condense the gas back into its liquid state. The liquid can then be brought into contact with the warm sea water and start the cycle all over again (Fig. 9-3). The sea water cools or heats the gas via heat exchangers. This system is a closed one, but there is also an open model where the sea water itself is the working fluid. Water is converted into a gas via a vacuum technique and the resulting steam, after driving the turbine, is returned to the ocean. For various reasons, including structural ones and the possibilities of salt buildup, the closed-cycle technique seems more promising.

The OTEC concept actually is fairly old having been first proposed in the 1880s by a French physicist named Jacques d'Arsonval. He suggested

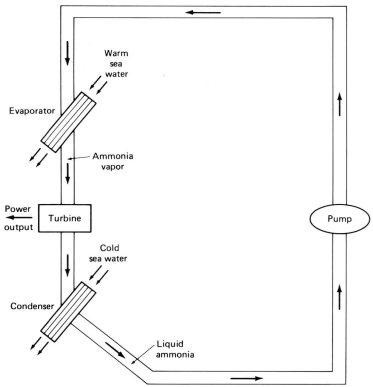

Figure 9-3.   Basic aspects of the OTEC system.

Figure 9-4. Model of the OTEC system proposed by TRW and Global Marine. One of the four generating units is shown. (Courtesy of TRW)

(d'Arsonval 1881) that the differences in ocean temperatures between the surface and deep waters could be used to produce electricity. A small plant, following this concept, was built in Cuba in 1930 but only lasted about 2 weeks before being destroyed by heavy seas. Following this, the idea languished until the mid-1960s, when it was picked up by Mr. Helbert Anderson and his son (Anderson and Anderson 1966) who considered it a viable method for producing cheap and continuous energy from the sun. The idea has received much more attention in recent years because of interest in the ocean as another source of energy. As late as 1972 the amount of funds available for OTEC from the U.S. National Science Foundation was less than $100 000, but by fiscal year 1977 it had climbed to about $14 000 000.

Essentially, two types of systems have been given principal consideration. One was proposed by TRW and Global Marine; the other by Lockheed Missile and Space Company. The TRW plan consists of a 340-ft (103-m) diameter concrete cylinder, the platform of which sits above the water (Fig. 9-4). It contains four power modules, with a central intake for cold water. The cold water pipe has a 50-ft (15-m) diameter and would extend down into the ocean to a depth of about 4 000 ft (about 1200 m) to reach the needed cold water. The temperature difference at the bottom of this pipe compared with surface water would be about 20°C, which should be sufficient to vaporize the working fluid in this system (probably ammonia) and to power the electrical generators.

Lockheed's design (Fig. 9-5 and 9-6) also uses ammonia as the fluid and likewise contains four power modules. It differs in that it is supposed to produce more power (187 million watts) and is to have an intake pipe made of concrete rather than the fiberglass that TRW is proposing. This pipe would extend only to a depth of 1500 ft (about 450 m). Estimated cost is $252 million or $13 350/kWh. Both systems will have to be connected to the sea floor.

Carnegie-Mellon University, the University of Massachusetts, and The Johns Hopkins University have also devised models for other types of OTEC systems. The Johns Hopkins model could make over 300 tons of

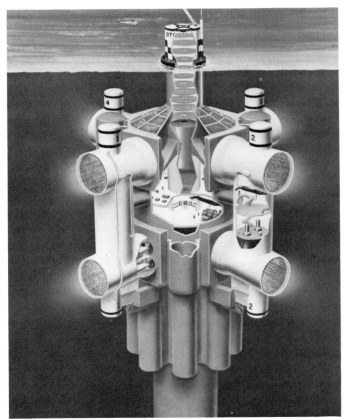

Figure 9-5.   The Lockheed Ocean Thermal Energy Conversion System, showing the major components. See Fig. 9-6 for details of operation. (Courtesy of Lockheed Missiles and Space Company, Inc.)

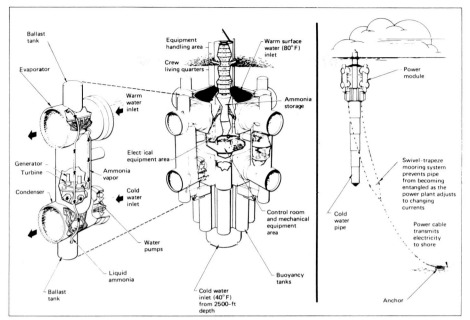

Figure 9-6. Cutaway figure of the Lockheed Ocean Thermal Energy Conversion System. The total length of the platform is 592 ft (about 180 m) and its diameter is 246 ft (about 75 m). Each of the power modules is 305 ft (93 m) high and 72 ft (22 m) in diameter. The cold-water pipe, which is used to collect the cold water, is 1000 ft (304.8 m) long (five sections). (Courtesy of Lockheed Missiles and Space Company, Inc.)

ammonia daily. It would be a floating plant that would also be able to produce ammonia, alumina, and other products. In 1978 a $25.4 million contract was awarded to TRW and Global Marine to design, build, and operate OTEC-1. This system will use a converted tanker for a platform to test a variety of heat-exchange concepts and will be operational by the early 1980s. It will be sited about 18 miles off Hawaii.

For an OTEC system to work, a temperature difference (surface to depth) of at least 30°F (16.6°C) should be available. This limits the locality where such facilities can be built without needing extremely long pipelines to the bottom. This is because there are not that many areas in the ocean where such temperature differences exist within a relatively short vertical distance. Because of the small temperature differential used and other losses in the OTEC system, efficiencies would probably be in the order of about 2%. This is relatively small when compared to a coal-powered generating plant, which may be as efficient as 30%. For an OTEC facility to be efficient, therefore, it must transfer about 10 times as much heat to yield the same power output as a coal plant. Any system built should be sited somewhat close to land or to the source of consumption of the energy. For the United States, it is estimated that the plant cannot be further than 100 miles off the shore. Among the more favorable localities are the Gulf Stream, Gulf of Mexico, and off such islands as Hawaii.

According to many, the main factor holding up the development of OTEC systems is money; others question the concept and doubt whether it ever would be effective. There certainly are several problems involved in an OTEC system; one is corrosion, which can damage or reduce the efficiency of the heat exchangers; marine organisms can also foul the system. Another problem is transmitting the energy from a plant at sea to the area of use on land. Environmentally, the OTEC system does not seem to be much of an issue; actually the environment seems likely to do more damage to the OTEC system than the system will do to the environment. Other problems include anchoring the system on the sea floor and the stresses that can develop on the cold-water intake pipe.

A U.S. National Academy of Sciences Panel[1] recently discussed many aspects of OTEC and whether it can ever actually be operationally competitive to other sources of energy. It specifically questions some of the optimistic costs and efficiencies. Part of the concern is that the system has such low efficiencies that a large amount of water would have to be used. For example, a plant producing 100 MW would use as much water as flows through Boulder Dam. To move this water around would require a significant portion of the energy produced by the system. Estimated costs are considerable, ranging as high as $2500 per kilowatt of output. Estimates, however, are probably premature because not enough data are available and all the technologic and potential construction problems are not known. One factor perhaps in favor of an OTEC system is the cost of other sources of energy. As this continues to rise the OTEC system will become more attractive.

Another debate concerning OTEC is whether the plant should be sited offshore or near shore. A near-shore location would make it easier to transmit the electricity via cables, whereas offshore plants could produce such chemicals as ammonia and hydrogen.

Within the next few years some critical decisions will be made concerning OTEC, especially as pilot plants are being built. Among the more optimistic estimates is that a major 100-MW plant could be built by 1985 and that by the year 2000 an important portion of the United States' electric supply could come from OTEC operations. In order to not overemphasize the value of OTEC, it should be understood that a 100-MW plant is only about $\frac{1}{10}$ the capacity of a modern fossil fuel plant or nuclear power plant. For OTEC it appears that the best way to obtain more power is to build more plants rather than larger plants. These plants, however, would have to be

[1]Panel headed by H. Sheets, University of Rhode Island; review conducted under the auspices of the Marine Board of the Assembly of Engineering of the National Academy of Sciences.

separated from each other by several kilometers to keep the surface cooling of the ocean to a minimum. Because OTEC plants will have to be sited in tropical areas for maximum efficiency, it is improbable that these facilities could ever be the principal source of energy for the world.

Artificial Upwelling

One aspect of many of the OTEC schemes is that the deep water raised to the surface will probably be enriched in nutrients and so could be used in an aquaculture or mariculture operation. If a food crop could be obtained from an OTEC system, it would reduce the total operating cost of the energy operation. Among the most active advocates of such a system have been Oswald Roels and his colleagues, who have built an experimental artificial upwelling system off St. Croix in the Virgin Islands (Roels and others 1973, 1975). The basic idea, which has other potential side benefits, is to bring cool, nutrient-rich water up from modest depths to the surface and use it to grow algae. The algae will grow at a much higher rate in the nutrient-rich water than in the nutrient-poor surface waters. These algae are then fed to animals higher up in the food chain, which, in turn, will also grow faster than they would under normal conditions.

The operation is based on the fact that large areas of the ocean have cold water at relatively shallow depths and that many islands, especially those with narrow shelves, are very close to these sources of cold water. In addition, many islands, especially those in the tropics or midlatitude regions, are situated below wind systems that carry large amounts of humid air. As part of a system the cold water could also be used to recover moisture from the atmosphere. This might be done by passing the sea water through a series of condensers placed in the flow of the moisture-laden winds. As the winds are cooled much of their moisture would condense out and could be recovered, stored, and used for drinking water or other uses. The cold water could also be used for cooling, perhaps for nuclear energy power plants or for large-scale "air conditioning systems" for the island. The total concept is extremely imaginative and may indeed present an interesting future use of the ocean. It does not seem to present any significant ecologic problems, except for the possible discharge of cold water into the relatively warm surface waters. The operation obviously is not without its shortcomings, one of which is that it would need a fairly substantial energy supply to raise the water. Roels and his colleagues recommend the interesting possibility of windmills.

The experimental system in St. Croix brought water from a depth of 870 m and stored it in large pools, where the cultures of planktonic algae are grown. This system daily produces over 100 000 liters of mainly diatom culture (with 10^4–10^6 cells per milliliter) which is then pumped into shellfish tanks (Roels and others 1975). The shellfish (species of oysters, littleneck clams, quahogs, and scallops) feed by filter feeding and remove up to 90% of the algae. In this manner, these organisms grow and reach market size considerably quicker than under normal conditions; some oysters have been ready for marketing within 6 months. The cold water, before it is used for mariculture, could be used for power generation, air conditioning, or desalinization (Fig. 9-7). The water can be warmed by heat from power stations, by standing in pools, and by its trip up the pipe. The cold water is also relatively free of disease-producing organisms, predators and parasites, fouling organisms, and anthropogenic pollutants, thus aiding the mariculture operation. After use, the water containing waste products from the animals is passed through a seaweed growing area and some wastes are removed. The seaweed itself can be a source of agar or carrageen. According to some proponents the potential yield from mariculture could be more valuable than the potential energy yield using temperature differences.

Energy from Salinity Differences

Power via salinity differences is another potential source of energy. The salinity differences between ocean water and river water is essentially maintained by the evaporation of sea water and by the rain received by rivers. In the ocean, salinity differences are generally very small, with surface waters usually being more saline than deeper waters; however, the difference may only be a few tenths of a part per thousand. The major salinity differences occur at or near the mouths of rivers where fresh river water comes in con-

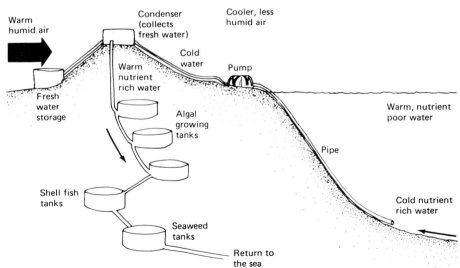

Figure 9-7 A possible schematic of an artificial upwelling mariculture system similar to that in St. Croix.

tact with sea water. At these areas energy may be obtained because of differences in osmotic pressure (caused by salinity differences) between the two waters. For example, the osmotic pressure differential between fresh water and sea water is equivalent to a head of water of about 240 m (or, for example, water in a dam of that height). Taking an extreme case the osmotic difference between the waters of the Great Salt Lake and fresh water is equivalent to a head of slightly over 4500 m. Obtaining this energy for use will involve some fairly complex technology and present some environmental difficulties. Nevertheless, the energy potential is significant. For example, 1 m^3 of fresh water dissolved into a large amount of sea water will dissipate about 2.25 MW of power (Weinstein and Leitz 1976), not all of which is recoverable. If the total world's river supply is considered, a theoretical supply of about 2×10^{12} W is possible, a value more than the world consumption of electricity, although little, and perhaps none, of this energy can be obtained. One possible method of getting energy from salinity differences is by dialytic batteries. A dialytic battery is composed of a series of alternating anion- and cation-exchange membranes forming separate compartments through which waters having different salt contents can flow. Electrodes attached to a source of power are at each end of the stack. The flow of elements across the membrane will form an electrical charge between the electrodes. Such a battery is technically feasible but is not presently economically practicable. Even under ideal conditions, these batteries could only provide a small amount of our energy needs (because of the large amount of water needed). On a large scale, these devices could produce environmental problems by eliminating or reducing salinity interfaces in estuaries and generating small amounts of hydrogen and chlorine. However, there would be no thermal discharge problems.

Energy from Marine Biomass

One of the more imaginative ideas for obtaining energy is to use marine biomass (i.e., plants or algae). Many marine plants can, under the right conditions, grow very quickly and after harvesting be converted into natural gas and other products.

One species commonly considered is the giant California kelp (*Macrocystis*), which grows as much as a foot a day and is easy to harvest. To increase the growth of kelp or other plants, artificial upwelling systems (or OTEC) can be used to provide increased nutrient supply. The kelp or other plant, when harvested, can be converted by either physical/chemical processes or biologic processes into natural gas. There also can be other by-products from the breakdown of the kelp that can be used for fertilizer or livestock feed (Fig. 9-8). General estimates indicate that the cost of gas, particularly methane, produced from biomass is about the same as that of gas produced from coal gassification processses.

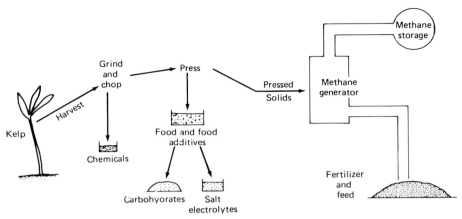

Figure 9-8. Conceptual model of a kelp processing plant showing the many possible by-products. (Adapted from Flowers and Bryce 1977)

A marine farm concept can be visualized that would produce fuel, food, and other products; it would be situated in fairly shallow water because the plants are attached to the bottom. Alternatively, an artificial device could be built if the "farm" is to be in deeper water, to allow attachment to a fixed device held above the bottom. In most instances extra nutrients will be needed, so this system would fit very well within an OTEC or artificial upwelling operation. Harvesting would be by cutting of the upper portions of the plants on a regular basis—similar to what is presently done in harvesting kelp for other uses.

Alternative systems that use microscopic algae also have been proposed. In these systems ponds or small lakes, perhaps using waste products, would be used to grow algae which will eventually be harvested and, by fermentation, produce methane. The residual nutrients from the process, particularly nitrogen and phosphorus, can be reused to feed the algae. If such a system were adjacent to or near an energy plant, the methane could be used to produce electricity and the unused heat that results from this process could be returned to the system to accelerate the growth of the algae in the ponds.

Some test facilities have been able to make conversions from sunlight to plants at efficiencies of about 4%–5% and then by using fermentation processes convert 65% of this material into combustible gases that contain as much as 50%–60% methane. Assuming a yield of algae of about 25 tons/acre/year, this could result in 300 000 000 Btu/acre/year in the form of combustible gas or if this was used to produce electricity 30 000 kWh/acre/year (Oswald 1974). Such a yield at this time is fairly close to being economically competitive with that of natural gas.

The concept of using solar energy via photosynthesis is reasonable because the sun is the basic source of energy on the earth. It is, of course, a renewable and essentially inexhaustible form of energy. Obviously as promising as solar energy is, it muct be emphasized that the efficiency of solar energy conversion to biologic plants is very small. These systems will not be a total solution for our energy problems but they may provide reasonable alternatives under certain unique conditions, such as where lakes, ponds, and land are available near existing or planned power facilities or by using parts of the coastal zone. If areas presently used for domestic organic sewage wastes were used, it is possible that a modest portion, perhaps 10% or more, of the present consumption of natural methane gas could be produced by this system.

Climate and the Ocean

A better understanding of climate could improve both chances of meeting future food shortages and our ability to predict or even to moderate the weather. In addition to the sometimes rather dramatic yearly changes in climate, there are also changes that extend over periods from tens of thousands to hundreds of thousands of years. The best examples of these are the recent glacial and interglacial periods. How these glaciations started is still a matter of conjecture, but probably the warming and cooling periods either have been influenced by or have affected (or both) the oceans in a rather dramatic way. Over shorter periods of time there have been fluctuating intervals of warm and cold periods. For example, from about 950 to 400 BC the weather was generally cool; it was relatively warm between 800 and 1200 AD and then cool again from 1550 to 1850. From about the 1900s on the weather has been relatively warm, although the cold winters of the last few years suggest that there may be a cooling period coming on.

There are clear interrelationships between the ocean and the atmosphere that are mainly due to the fact that most of the energy from the sun reaches the oceans, because the oceans cover 72% of the earth's surface. Because of the great depth and size of the oceans, they can absorb larger amounts of heat than the land. The incoming solar energy is absorbed in the upper layers of the ocean and heat and water vapor are returned to the atmosphere. Convective movements are initiated in the tropics (see Figs. 2-11 and 2-12), which in turn help drive the atmospheric circulation. Some of the incoming solar energy also drives the circulation of the deeper parts of the ocean (see page 19). The interactions of the ocean and the atmosphere are complicated by the fact that the northern and southern portions have large areas of ice which do not absorb heat at the same rate as the ocean.

Oceanographic and meteorologic research over the last couple of years has shown that the temperature of the ocean, especially in its upper few hundred meters, can vary from time to time and from place to place. These variations may be both the cause and perhaps the result of variable and changing weather conditions over extremely large areas. By processes not really understood, these conditions in the oceans may result in disturbed situations in the weather in areas thousands of miles downwind. These situations may also change the major and prevailing wind systems of the ocean into abnormal patterns that can produce equally abnormal weather conditions. These anomalous areas in the ocean, besides being large, are fairly deep and may have long-term effects on the weather that can last even for years or perhaps decades.

Changes in surface ocean temperature do not necessarily directly follow atmospheric changes but may lag 2 or 3 years behind them. In addition there can be some feedback and in this way the initial atmospheric change can be reinforced by what happens in the ocean. Some anomalous situations, such as those that occurred in the 1960s in the northeastern United States, where the summers were extremely dry, may have been caused by sea temperature anomalies within the ocean, although very little is known about how this may have occurred. There are also slow changes in the temperature of the deep layers of the ocean that may have unknown climatic effects and may even cause some of the longer term temperature changes mentioned above.

A major program, Global Atmospheric Research Program (GARP), was started in the mid-1960s to try to understand the changing conditions of the atmosphere and how the affect the weather. Other objectives were to develop longer forecasting periods and to understand the factors that determine the statistical properties of the circulation of the atmosphere. In broad terms these objectives are definitely related to interactions between the ocean and the atmosphere.

More recent experiments have shown that there are complex patterns within the ocean that are expressed as large-scale gravity waves, internal motions, variations in such currents as the Gulf Stream, formation of rings or spinoff of portions of water, and the previously mentioned changes in ocean temperature. The reasons for these, in many instances, are not really known.

A recent program, Mid-Ocean Dynamics Experiments (MODE), made a large series of various physical oceanography measurements in a 600-km diameter area of the Atlantic to try to detect major variations in the oceanic circulation pattern. In particular they hoped to observe (and succeeded) an ocean "eddy"—a slow-moving, high-energy colder or warmer core of a body of water moving through the ocean. These bodies are analogous to atmospheric high- or low-pressure systems, although how or whether the ocean counterparts influence climate is not yet understood.

The MODE program has been followed by one called POLYMODE, which is a combination with a similar Soviet experiment called POLYGON. There are many key questions concerning these eddies. Among them are how they affect the ocean's heat budget, influence currents, and affect climate.

One of the more interesting and challenging problems facing scientists in the coming years is to learn better how to predict and perhaps even control or modify the earth's weather. This challenge became obvious with the recent oil embargo, when it was important to know if the weather for the coming winter would be hard because there might be serious fuel shortages. Some large-scale oceanographic studies have shown that by studying the ocean one may be able to better predict changes in the earth's weather. A program called North Pacific Experiment (NORPAX) noted that there were shifts and changes in the ocean circulation and temperature patterns which seemed to have a distinct effect on weather. One of these is El Niño, which is an occasional and temporary stopping of the upwelling of water off the Peruvian coast. These waters are nutrient-rich and support a very large anchovy fishery. When the upwelling stops it triggers a series of minor catastrophes, including a drop in anchovy catch and the slowing of the guano industry (bird droppings used for fertilizer).

It has been found that the El Niño occurrences are caused by local changes in currents, wind patterns, and rainfall in the Pacific region. It may be that similar variations in temperature found in the Pacific are influencing changes in weather patterns over the United States. The changes in temperature are relatively small, being on the order of 1°–2°C above the average, but they cover large areas and may represent a major source or sink of thermal energy that could affect weather patterns. It should be emphasized, however, that the mechanisms involved are not understood; neither is it clear that the changes in the ocean are a response to an atmospheric change rather than the other way around. By studying these temperature changes over a long period of time and comparing them with weather on land, eventually some sort of predictive model could be developed which would allow long-range forecasting. One such analysis has been made by Dr. Namias (*Science*, 1974) of Scripps Institution of Oceanography who, working with other scientists in the NORPAX Project, has noted that these long-term temperature changes, which cover areas as large as 1000 km or more, tend to have an interesting pattern. For example, a cold patch would be found in one area and a warm patch in another. Dr. Namias feels that he has found correlations between oceanic and atmospheric data whereby the wind patterns are correlated or associated with the sea-surface temperature patterns. For example, cold, polar air may be sweeping down on a particular area and picking up heat from a relatively warm area.

One of the more interesting results of the NORPAX work is that examination of data collected over a long period of time has shown that in six relatively extreme years of climate there has been a distinct relationship between the sea-surface temperature pattern and the winter weather pattern over the United States (*Science,* 1974). When a warm patch of water was found off the California coast and a cold patch was found in the mid-Pacific, there was a generally colder than normal winter in the eastern United States and a warmer than usual winter in the western half of the country. Interestingly enough, the reverse pattern, i.e., abnormally warmer water in the mid-Pacific and mild winter weather in the eastern United States, was also found. The fact that only six such relationships have been found may not prove the point but it does suggest a link between oceanic temperatures and climate.

The results of the NORPAX program, although admittedly not yet conclusive, are a major step forward and emphasize the need for large-scale integrated research efforts involving many institutions and scientists from different disciplines using surface ships, stationary buoys, satellites, airplanes, etc. Such a major effort is beginning to be formulated in the United States and a large-scale climate program has been developed under the leadership of the National Oceanic and Atmospheric Administration (NOAA).

Hurricanes are among the most dramatic storms on the surface of the earth. Single hurricanes have caused damage in excess of a billion dollars and resulted in hundreds of lives being lost. When such a hurricane hits a coast it is possible that the winds associated with it may raise the sea level 10 m or more above average. Mechanisms for controlling and altering hurricanes obviously can have great economic value. One such program, called Project STORM FURY, created in cooperation with the U.S. Navy and NOAA, has performed experiments on hurricanes in 1961, 1963, and 1969. Perhaps the most impressive one was in 1969, when STORM FURY scientists seeded a hurricane with silver iodide crystals. This would cause super-cooled water drops to freeze and release their latent heat. The result should be to heat up the hurricane. On the 18th and 20th of August such seedings were done; they resulted in a decrease of wind velocity of 31% on the 18th and 15% on the 20th. No experiment was done on the 19th, and on this date the storm regained its original intensity. The results of the experiment are equivocal because such changes as this can happen without any seeding, but they suggest the possibility of adjusting or modifying the effects of hurricanes or large storms.

It should be emphasized that modifying oceanographic or weather conditions could produce unanticipated dramatic changes. For example, plans to modify or adjust weather, in particular in the polar regions, have always been appealing for the Soviet Union because a large number of their ports are ice bound for large parts of the year. The idea of controlling the weather and the climate in the Arctic is not really a very new one. In the latter part of the eighteenth century, Nansen had proposed the idea that if one were to increase the cross-sectional area of the Bering Strait, there would be the possibility of increased inflow of warm Kuroshio Current water into the polar basin; this would decrease the ice cover and perhaps make the climate less violent. More recent ideas have been to cover the ice with dust or dark powder to help it melt, to get rid of the cloud cover which should increase melting of the ice, and to move or redirect Arctic water into the Kara Sea.

Fresh Water from the Sea

In future years, besides energy and food problems, certain countries and regions will face critical shortages in fresh water. Even now many un-developed countries do not have enough water to maintain even their presently low standards of living. Two obvious places to look for fresh (or sweet, as it is sometimes called) water are the oceans or the ice caps and glaciers. These two environments contain 97.2% and 2.15% of the world's water supply, respectively, whereas all the water in freshwater lakes, rivers, and streams totals about 0.01% (see Table 2-3).

The problem of obtaining fresh water should really be considered in two ways. The amount of water needed for drinking and domestic use is not very large and most countries or individuals will pay a reasonable premium for it. This water has to be fairly pure and always available. The amounts of water needed for agriculture and industry, in contrast, are considerable compared to human consumption, but the water does not have to be as pure. In fact, as we shall see in a few paragraphs, some crops may be able to grow in brackish or saline water.

There are three major problems in converting sea water for aquaculture usage: First, a major source of energy has to be available; second, a process must be used to remove the salt from the water; and third, transportation mechanisms have to be available to move the water economically. Desalinization is the basic technique for removing salts from sea water. Desalinization plants, however, will produce water at a fairly constant rate, and in agricultural usage there will be peak periods of demand followed by periods when water is not needed; therefore, storage facilities or other uses of the water are necessary.

The desalinization of sea water, although it is a complex technology, is usually based on a fairly simple process. When sea water is turned into a vapor, the salts remain behind. The vapor, therefore, is composed of fresh water and, when condensed and collected, can be used for drinking. This evaporation process is one of the most common methods of desalinization.

There are six basic methods of desalinization that are presently known: freezing, humidification, distillation, absorption, electrodialysis, and reverse osmosis. Freezing is based simply on the fact that when sea water freezes the ice that is formed is pure water, whereas the salts are left behind in a con-centrated brine. The energy and technology for this operation compare favorably to those of other techniques, but it takes longer to freeze water than to form steam or to produce water by evaporation. Likewise, it is a little more difficult to handle water in the frozen state than in the fluid or gaseous state. Another problem is that water in the frozen state will occupy considerably more space than water vapor; therefore large areas are needed (the use of water that is already frozen is discussed later in this section).

Humidification, a technique based on evaporation, is fairly simple and can use solar energy. Large pools of water enclosed within glass areas can be heated by the sun; the water evaporates and condenses on the glass and then can be collected and used as a source of fresh water. The amount of water collected in this manner will not be very large; nevertheless, it is a relatively cheap method that can be used in local areas; the energy costs are fairly low. Distillation, in its simple form, involves boiling salt water and cooling the vapor to collect fresh water; the salt is left behind in a concentrated brine. There are sophisticated methods, including a multistage flask (using a vacuum) distillation technique. Among the problems with this technique are the formation of large amounts of salt, or scale, that have to be removed from the evaporating area and the need for a fairly large amount of energy.

The absorption technique is simply a method in which the salts are removed from sea water by some sort of chemical or filtering agent, which can be a resin or activated charcoals. These systems can be constructed in small kits for use on rafts and in local situations.

Electrodialysis is a technique that uses semipermeable membranes, which permit the ions in the salt water to pass through but not the water itself. In the process sea water is separated from fresh water by membranes, electrodes are put into the freshwater side, and when the current is started positive ions will flow from the sea water through the membranes toward the negative electrode, whereas negative ions will move toward the positive electrode. In this manner ions are eventually removed from the salt water until it becomes fresh. Some technology improvements are necessary before this technique is compatible with other methods. One of the more promising techniques is the reverse osmosis method. Osmosis, a natural process, is the movement of less salty water toward more salty water through a semipermeable membrane. By reversing the process (by using pressure) it is possible to force the water to flow from a salty solution to a less salty solution (Fig. 9-9). The forcing will move the water through the membrane but leave the salts behind, thus increasing the salt concentration in the salt water but producing fresh water on the other side of the membrane. This technique can also be used in the treatment of sewage. One of the present problems in this process is that the membranes do not last very long.

Osmosis units are especially effective in converting brackish water into potable water, and presently units are being used in at least 28 countries. This technique has become especially important in the Middle East, where economic growth has lead to a doubling in such units over the past 3 years. In this system and others there is considerable advantage to using brackish

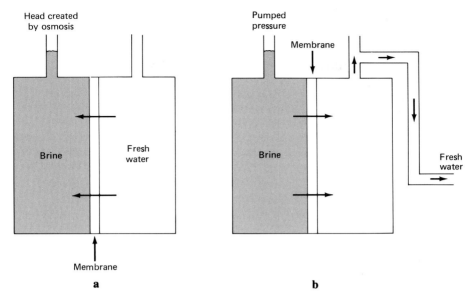

Figure 9-9, a and b. **a** Natural osmosis and **b** reverse osmosis. In the natural system the fresh water causes a hydrostatic head, making the fresh water flow into the brine. In the reverse osmosis system a pump drives the fresh water out of the brine and through the membrane.

water rather than salt water because the former contains less salts and impurities and the filters can last longer. Among some of the advantages of the reverse osmosis process is that corrosion can be kept to a minimum, because many of the operating parts can be plastic. Scale concentration is also generally low or absent, and the units can be small and have low energy costs.

In the flask evaporation technique it is possible to use heat from an electrical or nuclear power plant that in many other situations would be wasted as a form of energy. The cost of producing water by this technique can be as low as 20¢/1000 liters but it is felt that these costs can be reduced even by half with much larger facilities. One of the largest such units is located in Kuwait and has a capacity of about 22 million liters per day.

At present, evaporation techniques are the best known and understood procedures. A problem with any of the systems is to make them economically competitive, which is not hard to do if the water is used only for drinking. In the case of agriculture and industry usage, however, cost becomes much more important because so much more water is needed to produce a relatively low-priced product. Actually there is considerable evidence that using desalinated water for agriculture may not really be economical except under very unique conditions.

An alternative to using desalinated water for agriculture is to develop farm products that can be grown in salty or brackish waters. Some recent experiments at the University of California (Epstein and Morlyn, 1977) have suggested that this may indeed be a possibility. Researchers there have developed barley that can grow in seawater irrigation rather than in fresh water. The logic of this approach is that it is probably easier to adapt the crops to salt conditions than to adapt the environment. Another advantage of this method is that large areas of California, and other places, have already been adversely affected by salt in the water and so, rather than abandoning these areas for agriculture, it is better to develop crops that can grow in salty or brackish water. If salt has been concentrated in a soil the general procedure is either to stop growing crops or to try to flush the salt out by using large amounts of fresh water.

The initial attempts to grow crops in salt solutions were not very successful, as most of the original plants died. However, the few that survived produced seeds and, by using these seeds it has been possible to grow plants that are salt resistant. The yields are about half that of freshwater crops. Similar experiments are also underway using salt-resistant tomatoes. It is neither obvious nor necessary that these techniques ever will replace freshwater crops, but the techniques do indicate the possibility, under certain controlled conditions, of using these plants as an alternative. Other products that may be applicable to sea water and have had some success in experiments are celery, sugar beets, swiss chard, asparagus, spinach, and some grains.

Another completely different possible source of fresh water is icebergs. Icebergs (the best come from Antarctica) would be towed by some mechanism to the areas where water is needed. The most publicized such area is Saudi Arabia. The credit for this idea probably belongs to Professor John Isaacs of the Scripps Institution of Oceanography, who suggested it informally several decades ago. The concept was given some recent enthusiasm because of a 1977 conference held in Iowa entitled First International Conference in Iceberg Utilization. The meeting was sponsored by a Saudi Arabian Price and brought together about 200 scientists and respresentatives from as many as 18 countries to discuss the possibilities, technical, environmental, and legal, of transporting and using Antarctic icebergs.

The iceberg is a fairly good source of water because it is quite pure—in most cases, purer than ordinary drinking water. Saudi Arabia is not the only country that may have use for such a large source of fresh water. Other possibilities include California and Australia.

Among the items discussed at the meeting was how, indeed, such an iceberg can be moved—for example, by towing it on the surface, having submersibles push it, or even putting propellers on it. A major problem is how to prevent the iceberg from melting during its trip (which could take a year). Wrapping it in plastic or spraying it with urethane foam were suggested. Another problem, of course, is the environmental aspect; what are the biologic effects when you tow an iceberg through another country's waters or exclusive economic zones—could it even affect the weather? Certainly the cold iceberg in some climates could produce fog or even rain. If indeed the iceberg could be transported to a needy area, additional problems would be how it actually would be handled once it is there. A complicated supply systems will be involved; portions of the iceberg will have to be broken off and moved, melted, etc.

The Saudi sponsor of the meeting has a company called Iceberg Transport International that is considering moving a 100 000 000-ton iceberg from Antarctica to the Arabian Peninsula. It would be wrapped in cloth and plastic and be towed by several ships. Their estimated trip time is approximately 8 months and will cost about $100 000 000. According to their calculations, if the mile-long iceberg lost 20% of its mass in travel, it would still cost less than desalinized water.

Other people, particularly oceanographers, have a hard time visualizing the iceberg losing that small an amount of water during the trip across the equator and some have stated that by the time you reach the equator you will be towing nothing but a rope. However, it may be possible to move an iceberg to other places that need water and are much closer, such as Australia or South America. Although this may sound like a frivolous and perhaps improbable use of the sea, the Saudi sponsor at the meeting noted that his country would spend $15 000 000 000 on water desalinization in the next few years.

Disposal of Nuclear Wastes in the Deep Sea

A major problem of the nuclear age is the disposal and management of the high-level waste products from the world's nuclear power plants. The volume of these long-lived and extremely toxic materials is constantly increasing and is expected to reach 13 000 m^3 by the year 2000. Basically, three management options exist:

1. Extraterrestrial transport and disposal (i.e., into space)
2. Transmutation or elimination of the dangerous components, by nuclear processes, into more tolerable isotopes
3. Safe disposal within the earth's environment

The technologies for the first two options are neither adequate nor dependable at this time, although they are very appealing. At this time the best option is disposal on earth, and there are four areas where this could be done. One is storage on land (or slightly buried) in strong and guarded containers. It will take up to 1 million years until nuclear waste products are safe; therefore it is necessary with this method that some society be willing to guard these facilities for this period of time. This is difficult to guarantee. A second possible disposable environment is to put the material in buried rock or salt formations. This idea was long the favorite one, but it has been found that many of these formations are not as geologically stable as anticipated and that high-level nuclear waste can cause dangerous chemical and physical reactions.

A third suggestion was to store the radioactive material in canisters within the world's major ice sheets. However, because climate conditions change (see earlier parts of this chapter) and ice may move, it seems dubious that the ice sheets would persist for the required time.

The fourth storage possibility, and the one that is getting much consideration at present, is below the sea floor of the ocean. There are two principal reasons why this environment is appealing. The first is that the sea floor is probably the least valued piece of real estate on earth. This is especially true of the deep sea, which has little probability of use for fishing or oil and gas production. The main use of such a region could be for manganese nodule mining but the size of the deep sea is so immense that plenty of room is available. Second, the deep sea is among the most stable parts of the earth. Some people have suggested that radioactive material could be stored in unstable areas, such as trenches. The logic for this is that trenches are zones of subduction (Fig. 2-8) and by this process the stored material would eventually become incorporated into the crustal material of the earth; a process that would take several millions of years. The concept is good for conventional waste products, but with the high incidence of faulting, slumping, and earthquake activity in trenches there is a good possibility of rupturing canisters placed there, with the possible escape of the material into the environment. The best environment in the ocean seems to be the central parts of the major ocean plates (Figure 2-9). These areas have extremely slow sedimentation rates (1 cm or even less per thousand years) and slow currents, are geologically very stable, and are unaffected by major climatic changes, such as ice ages.

In spite of the many advantages of the deep sea for disposal there still are numerous scientific and technical questions that must be answered. Among the important problems are:

1. Will the deep-sea sediments form a sufficient protective barrier to any escaping material and can they tolerate the heat that is generated by the radioactive decay?

2. What methods and procedures should be used to emplace the material and what should be their spacing? Some of the possible methods are shown in Fig. 9-10.

A valuable aspect to using the ocean is that the sediments, and perhaps even the water itself, can be part of the containment system (Fig. 9-11). Deep-sea sediments appear to have sufficiently high sorption capabilities to

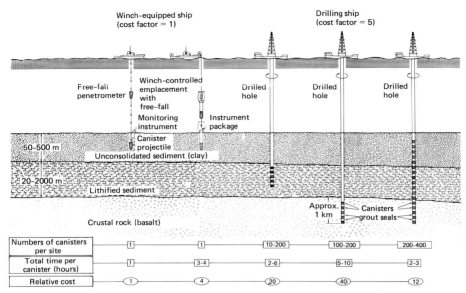

Figure 9-10. Various engineering concepts for emplacement of radioactive waste canisters in the sea bed. (Silva 1977)

prevent most radionuclides that may escape their canisters from reaching the ocean. In addition, the sediments have very low permeabilities, which reduces the possibilities of flow of material. These two factors by themselves are thought by some scientists to be effective enough to act as a barrier for several million years.

The areas where the canisters may be located are presently outside of any national jurisdiction in the ocean (see Chapt. 3). However, potential regulations resulting from the Law of the Sea Conferences could change this situation.

Innovative Opportunities for Using Marine Organisms

The animals and plants of the sea are clearly different in many aspects from those on land. These differences present some interesting challenges; for example, when one fishes one is essentially acting as a hunter, whereas if one could find a way of herding or attracting fish, one's catch per unit effort would improve. One mechanism to attract fish is to use sound and lure them into a net or some other harvesting device. This device does not have to be very sophisticated. For example, fish schools have been herded for centuries by fishermen making underwater sounds by banging poles or stones together. It seems reasonable that electronic devices that simulate specific

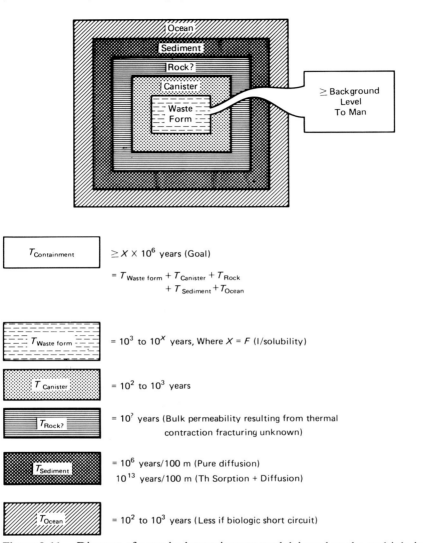

Figure 9-11. Diagram of a sea-bed containment model, based on the multiple-barrier concept, for radioactive wastes emplaced into the sea bed. Note that the sediment itself forms the most long-lasting barrier. (Hollister 1977)

sounds, such as fish eating, would work. The Japanese have been pursuing this type of research fairly intensely and have had some success with luring schools of fish by acoustic devices. Alternatively, sound could be used to scare fish away, for example, in keeping porpoises away when tuna are being caught or discouraging sharks from entering swimming areas.

Another rather interesting experiment concerns salmon, wherein chemicals are used to influence the fish to return to specific areas. The experiment was done using juvenile coho salmon, which were exposed to two different chemicals for about 1½ months and then released into Lake Michigan. Eighteen months later, when the spawning migration began, these same two chemicals were released into two separate streams. The results showed that the majority of fish exposed to the first chemical returned to the stream where the first chemical was subsequently placed. Likewise, most fish exposed to the second chemical returned to the site where the second chemical was being introduced. The experiment (Scholz and others 1976) shows that coho salmon can be imprinted with a particular chemical and made to return to any area where this chemical is introduced at a later time. Such techniques could be useful in breeding fish and having them return when they have reached edible size.

Some species of marine animals appear to be especially intelligent, especially such mammals as porpoises and whales. The ease with which porpoises can be trained to do rather remarkable stunts attests to this fact. Whether these animals can ever be trained to aid in catching or herding fish remains to be determined. Whales and porpoises, however, have been trained to recover such devices as torpedos from the sea floor. The U.S. Navy has whales that will dive to depths of more than 500 m to retrieve such items. One animal used, a pilot whale, could lift an object weighing 600 lb (about 270 kg) from 300 m or 300 lb (about 135 kg) from 600 m (these weights are submerged weights). Other animals, such as sea lions and porpoises, have also performed such feats.

Perhaps in future years the most important use for the animals in the sea is that they are a source for new and valuable drugs. Obtaining drugs from the sea has been done for centuries; for example, the Japanese and Chinese ate seaweed which was a source of iodine and apparently was responsible for their low incidence of goiters. There is a broad folklore concerning the use of substances from the ocean for medicine, both for preventive and for healing aspects. For example, Pliny the Elder recommended using cinders from the burnt spine of a stingray with vinegar to relieve toothaches (Ruggieri 1976).

The systematic search for drugs from organisms in the sea, however, is a fairly modern phenomenon, which began in earnest in the late 1950s or early 1960s. Because many marine organisms are toxic, they have the potential for drugs, for poison indicates strong potential physiologic activity. At present, about 2000 organisms in the marine kingdom are known to be poisonous or venomous. Other possible sources of drugs may result from the fact that some species have profound effects upon others. One good example is the so-called red tide that results from a rapid growth of dinoflagellates, a microscopic plankton, which can cause fish kills. The exact reason for the increase in growth is not understood but when it happens it can actually color the water and cause a large mortality among fish either because of the toxins from the plankton or because of the removal of oxygen by the activities of the increased numbers of plankton, or both. Although the fish die, shellfish organisms that eat the phytoplankton do not; instead, they accumulate the poison. This, in turn, can be dangerous and perhaps even fatal for humans who may eat these shellfish.

The successful use of toxic chemicals as a source of medicine is exemplified by the sulfur and penicillin drugs developed in the early twentieth century. This led to the increased search for new and exciting drugs among marine organisms, in particular among animals that may have poisonous or toxic compounds. The diversity of marine life also makes it an especially attractive place to look, because many land species have already been examined. Actually little is known about many marine species. For example, it has been recently found that some marine organisms contain antibacterial agents that prevent the growth of the bacteria in their intestinal areas.

Many varieties of algae and seaweed are used in the food, cosmetic, and drug industries. The brown algae can yield alginates, whereas red algae produce agar and carrageenan. Calcium alginate, when incorporated into surgical dressings, can reduce bleeding and is also used in various forms in dentistry. Agar is used in the growth of bacteria in culture plates. Besides algae, other animal species, such as sponges, jellyfish, worms, snails, sea cucumbers, fish, and horseshoe crabs, can be used to obtain various drugs. One fairly popular medicine is cod liver oil, which is a source of vitamins A and D.

The finding of new drugs often is a case of serendipity. For example, cephalothin, a relatively new drug useful against germs resistant to penicillin, was found in an Italian sewer outlet. To find these new drugs requires almost a systematic examination of each different animal and plant species to determine its composition and what the potential drugs may be. Essentially, the process involves collecting different organisms and testing them for their various properties. Any potential drug must, of course, be tested on animals to determine its effects and any unwanted side effects before it can be used on humans. Some of the more interesting recent finds have been that material from some sponges has shown strong antibiotic capabilities; poisons from fish appear to be useful in reducing blood pressure and some extracts have slowed or reduced tumor growth in experimental animals. One of the major efforts has been to find an anticancer drug; cancer seems to be relatively rare in many marine animals.

It may also be possible in the future to use marine organisms to extract selected elements from sea water. Many marine animals will concentrate trace elements from the sea water into their skeletons or tissues by factors of over 10000 times the concentration of the element in sea water. If the right animals or plants could be cultivated they could provide a source of certain valuable elements by extracting them from the sea water.

Other Innovative Ideas

Satellites

A technology that holds great promise for oceanography is satellites. One of the more important systems is the Earth Resource Technology Satellites (ERTS) of the National Aeronautics and Space Administration (NASA). (Recently the name of this system was changed to LANDSAT.) The orbit of these satellites was especially chosen to provide a systematic and repetitious coverage of the land and ocean. They revolve around the earth essentially every 103 min in a circular, almost polar, orbit about 800 km high. The satellite will make 14 revolutions around the earth each day, photographing three strips of about 185-km width in North America and 11 strips over the rest of the world. Each strip, from one day to the other, will be contiguous to that of the previous day with a small overlap; therefore, every 18 days it will pass over every location of the earth at essentially the same time and with similar lighting. This repetitive coverage can be very valuable in monitoring changes over time. A new satellite, SEASAT, was launched in 1978 especially for oceanographic observations. Initially it produced some remarkable pictures and data; unfortunately, however, it stopped working a few months after its launch.

Pictures from these satellites can be useful for monitoring ocean dumping and oil pollution (Fig. 9-12), prospecting for minerals on land, mapping coastal zone erosion, locating schools of fish, and other possibilities. Water pollution usually can be easily detected from satellite pictures, as can algal blooms in lakes and sometimes oil spills. In many instances color-enhanced photographs will yield even more data and can be used to distinguish subtle differences. The repetition of the observations may be valuable in studying the distribution and movement of suspended sediment (Fig. 9-13) and effluent from industrial plants and in detecting upwelling areas. Likewise, coastal zone erosion and deposition can be monitored.

Satellites and airplanes can also be used to collect routine data, such as sea-surface temperature and such information as sea state and weather; these can be transmitted worldwide and be of value for shipping (courses can be adjusted to avoid areas of bad weather). Both devices are also used to locate and trace the movement of icebergs, which is important for navigational purposes. Areas of high biologic productivity and sometimes even large schools of fish can be observed by changes in water color or perhaps even by the detecting of films of fish oil or algae accumulations. With some satellites, shallow bathymetry can be ascertained. It is possible, using appropriate camera techniques, to measure bottom topography (Fig. 9-14) to depths as great as 35 m, depending on sea state and water turbidity. This information, combined with changes in coastal features and observations of

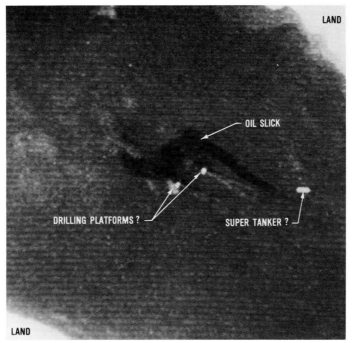

Figure 9-12.   LANDSAT photograph made over the Gulf of Suez, showing an oil slick and other objects. (Photograph courtesy of NASA)

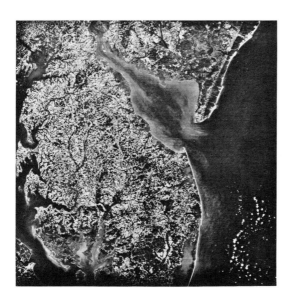

Figure   9-13.   ERTS photograph showing sedimentation patterns in Delaware Bay, (see Fig. 8-2). (Photograph courtesy of NASA)

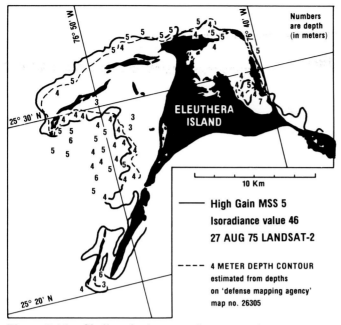

Figure 9-14. Shallow bathymetry in an area in the Bahamas as determined from LANDSAT observations. Note the close correlation with depths determined from conventional mapping techniques. (Courtesy of NASA)

movements of river effluents and sediments, can be very useful in evaluating coastal processes.

The measurements of sea-surface temperature from an airplane or a satellite will cover a large area at a single time, whereas a research ship makes one measurement at a time and then travels to the next spot at a speed of about 10 knots. These measurements from the air are not, of course, as accurate as those by conventional shipboard methods, but they still can be very useful, for example, in monitoring the discharge of hot water from nuclear or power plants sited in the coastal zone. Airplanes could also be used, as well as perhaps even satellites in some cases, to detect submarines.

Offshore Islands or Platforms

Human beings, for various reasons, have always been fascinated with living offshore on artificial islands or even underneath the sea (Fig. 9-15). The technology required for living underwater is available, but for each person staying underwater several will be needed on the surface to provide logis-

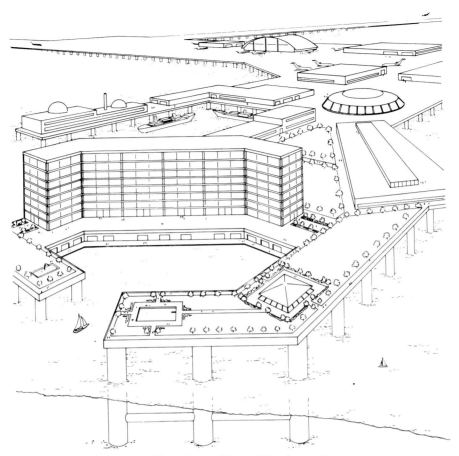

Figure 9-15.　Artist's rendition of an offshore island complex.

tical support. The idea of using the sea floor for habitats once we run out of space on land is just as viable as sending people to the moon or other planets for the same reason. Casual living either offshore or underwater, because of both its cost and its inconvenience, will never replace living on land, but surface or undersea habitats can be valuable in certain uses of the ocean. There are situations where an offshore facility would be a better use of the environment than having the same facility on land. Possibilities include offshore nuclear power stations, offshore airplane-landing strips, drilling platforms, fish processing factories, and recreational or scientific facilities. One major requirement of any such facility is that it be protected from the destructive forces of the waves.

Offshore nuclear power plants do have several advantages over land-based ones. Perhaps the key one is that the ocean will provide a more than adequate source of cooling water, and that when this now heated water is returned to the environment the potential damage from thermal pollution will be reduced because of the larger area into which it can be dispersed. In addition, by offshore dispersal, the water should not directly affect the coastal biologic environment. The offshore setting could make nuclear power plants more acceptable to their opponents because of their distance from large population centers. On the negative side would be the high cost of such plants and their possible damage by the marine environment (very few things survive indefinitely in the ocean).

Offshore airports or extensions of landing strips are a common idea and in many localities have already been built. The generally nearby source of sand on the shelf that may be available for construction also makes such a facility appealing. In many localities there are no alternatives to offshore extensions because the present airports are already constructed in highly industrialized and valuable land.

The use of offshore islands for drilling or loading terminals is common (Figs. 4-7 and 4-8), and such loading facilities reduce near-shore pollution and eliminate extensive dredging if the onshore facilities are not adequate. Drilling platforms number in the thousands and are especially common in such areas as Lake Maracaibo, Venezuela and the Gulf of Mexico.

Artificial reefs are another form of offshore island but these are underwater. Such reefs can be composed of ships (either deliberately or accidently sunk) or such items as old cars, rocks, or shells. Such habitats generally will attract large numbers of fish. In Hawaii, an artificial reef built from old cars increased the standing crop of fish from less than 40 lb/acre to over 1500 lb/acre. Streetcars dumped off California and beer cases off New York have had similar success. More imaginative use of these facilities could be a real boon to the commercial and sports fisherman. However, ecologic and oceanographic studies should be made before such a site is developed. For example, with unfavorable currents it is possible for the dumped material to end up on the adjacent beach area or, alternatively, to be rapidly buried on the bottom. The reef should also be placed in an area where the chances of attracting the fish are good. The material used must be considered, because corrosion can create an unfavorable environment. A side benefit of using "junk" for such a habitat is that it removes it from the rapidly limited space on land. However, just indiscriminate dumping at sea is not a sensible choice.

The preceding paragraphs have described some of the possible innovative uses of the sea. The vastness of the ocean and its many unique physical characteristics suggest that even more new and imaginative ideas are likely to be developed. These new concepts, as well as the ones described, will require considerable policy and scientific investigations so that the full effect of these uses can be evaluated.

References

*Anderson, J.H., and J.H. Anderson Jr., 1966, Thermal power from seawater: *Mech. Eng.,* v. 88, p. 41–46.

Angino, E.E., 1977, High-level and long-lived radioactive waste disposal: *Science,* v. 198, p. 885–890.

Bascom, W., 1969, Technology and the ocean: *Sci. Amer.,* v. 221, p. 199–217.

Bishop, W.P., and C.D. Hollister, 1974, Seabed disposal—Where to look: *Nuclear Technol.,* v. 24, p. 425–443.

Booda, L.L., 1974, An ocean based solar-to-hydrogen energy conversion concept: *Sea Technol.,* p. 21–24.

Bostrom, R.C., and M.A. Sherif, 1970, Disposal of waste material in tectonic sinks: *Nature,* v. 228, p. 154–156.

Broecker, W.S., 1975, Climatic change: Are we on the brink of a pronounced global warming? *Science,* v. 189, p. 460–463.

Brown, C.E., and L. Wechsler, 1975, Engineering an open cycle power plant for extracting solar energy from the sea: *In:* Dugger, G.L., Ed., *Proceedings, Third Workshop on Ocean Thermal Energy Conversion (OTEC).* Houston: Johns Hopkins University, Applied Physics Laboratory APL/JHU SR 75-2, p. 75–78.

Clampitt, B.H., and F.E. Kiviat, 1976, Energy recovery from saline water by means of electrochemical cells: *Science,* v. 194, p. 719–720.

Charlier, R.H., 1968, Tidal power: *Oceanol. Internatl.,* September/October, 1968, p. 32–35.

Claude, G., 1930, Power from the tropical seas: *Mech. Eng.,* v. 52, p. 1039–1044.

Clawson, M., H.H. Landsberg, and L.T. Alexander, 1969, Desalted seawater for agriculture: Is it economic? *Science,* v. 164, p. 1141–1148.

Cohen, R., 1975, Ocean thermal energy conversion: *In:* Morgan, M.G., Ed., *Energy and Man: Technical and Social Aspects of Energy.* New York: Institute of Electrical and Electronics Engineers, p. 218–220.

Craven, J.P., 1975, Present and future uses of floating platforms: *Oceansus,* v. 19, p. 67–71.

Damon, P.E., and S.M. Kunen, 1976, Global Cooling? *Science,* v. 193, p. 447–453.

*d'Arsonval, J., 1881, Utilisation des forces naturelles: Avenir de l'ectricite: *Rev. Sci.,* pp. 370–372.

Deese, D.A., 1977, Seabed emplacement and political reality: *Oceanus,* v. 20, p. 47–63.

deMarsily, G., E. Ledoux, A. Barbreau, and J. Margat, 1977, Nuclear waste disposal: Can the geologist guarantee isolation? *Science,* v. 197, p. 519–527.

Douglass, R.H. Jr., J.S. Hollett, and A.J. Karalis, 1975, Ocean thermal energy: An engineering evaluation: *Offshore Technol. Conf. Proc.,* v. 2, p. 22–38.

Dowling, J.E., and H. Ripps, 1976, From sea to sight: *Oceanus,* v. 19, p. 28–33.

Epstein, E., and J.D. Norlyn, 1977, Seawater-based crop production: A feasibility study: *Science,* v. 197, p. 249–251.

Floating nuclear plants: Power from the assembly line: 1974, *Science,* v. 183, p. 1063–1065.

*References indicated with an asterisk are cited in the text; others are of a general nature.

*Flowers, A., and A.J. Bryce, 1977, Energy from marine biomass: *Sea Technol.*, Oct, p. 18–21.

Friedman, W., 1971, *The Future of the Oceans*. New York: George Braziller, 128 pp.

Frosch, R.A., 1977, Disposing of high-level radioactive waste: *Oceanus*, v. 20, p. 4–17.

Gerard, R.D., and O.A. Roels, 1970, Deep ocean water as a resource for combined mariculture, power and fresh water production: *Marine Technol. Soc. Jour.*, v. 4, p. 69–78.

Gerard, R.D., and L. Worzel, 1972, *Atmospheric Moisture Extraction over the Oceans in: Biological Modification of the Marine Environment*. Washington, D.C.: National Academy of Sciences, p. 66–77.

Gilman, D.C., 1974, Feasibility of utilizing submerged turbines for extracting power from the Flordia Current, considerations for a feasibility study and pilot installation: *In:* Steward, H.B. Jr., Ed., *Proceedings of the MacArthur Workshop on the Feasibility of Extracting Useable Energy from the Florida Current*. Miami: NOAA Atlantic Oceanographic and Meteorological Laboratories, p. 306–321.

Golay, M.W., 1974, Offshore nuclear power stations: *Oceanus*, v. 17, p. 46–52.

Golueke, C.G., and W.J. Oswald, 1963, Power from solar energy—Via algae-produced methane: *Solar Energy*, v. 7, p. 86–92.

Goss, W.P., W.E. Heronemus, P.A. Mangarella, and J.G. McGowan, 1975, Summary of University of Massachusetts research on Gulf Stream based ocean thermal power plants: *In:* Dugger, G.L., Ed., *Proceedings, Third Workshop on Ocean Thermal Energy Conversion (OTEC)*. Houston: Johns Hopkins University, Applied Physics Laboratory, APL/JHU SE 75-2, p. 51–63.

Grant, P.T., and A.M. Mackie, 1977, Drugs from the sea—Fact or fancy? *Nature*, v. 267, p. 786–788.

Gray, T.J., and O.K. Gashus, Eds., 1972, *Tidal Power*. New York–London: Plenum Press, 630 pp.

Griffin, O.M., 1977, Power from the oceans' thermal gradients: *Sea Technol.*, August, p. 11–40.

Haber, G., 1977, Solar power from the oceans: *New Sci.*, 10, March p. 576–578.

*Hammon, A.L., 1974, Long-range weather forecasting: Sea temperature anomalies: *Science*, v. 184, p. 1064–1065.

Heath, G.R., 1977, Barriers to radioactive waste migration: *Oceanus*, v. 20, p. 26–30.

Heronemus, W.E., 1974a, Using two renewables: *Oceanus*, v. 17, p. 20–27.

Heronemus, W.E., 1974b, Alternate energy sources from the ocean: *Marine Technology Society Jour.*, v. 8, p. 35–38.

Heronemus, W.E., P.A. Mangarella, R.A. McPherson, and D.L. Ewing, 1974, On the extraction of kinetic energy from oceanic and tidal river currents: *In:* Stewart, H.B., Jr., Ed., *Proceedings of the MacArthur Workshop on the Feasibility of Extracting Useable Energy from the Florida Current*. Miami: NOAA Atlantic Oceanographic and Meteorological Laboratories, p. 138–201.

Hessler, R.R., and P.A. Jumars, 1977, Abyssal communities and radioactive waste disposal: *Oceanus*, v. 20, p. 41–46.

*Hollister, C.E., 1977, The seabed option: *Oceanus*, v. 20, p. 18–25.

*Inman, D.L., and B.M. Brush, 1973, The coastal challenge: *Science*, v. 181, p. 20–31.

Isaacs, J.D., 1973, The ocean margins: *In* English, T.S., Ed., *Ocean Resources and Public Policy*. Seattle: Univ. Washington Press, p. 79–93.

Isaacs, J.D., and R.J. Seymour, 1973, The ocean as a power resource: *Internatl. Jour. Environ. Studies*, v. 4, p. 201–205.

*Isaacs, J.D., D. Castel, and G.L. Wick, 1976, Utilization of the energy in ocean waves: *Ocean Eng.*, vol. 3, p. 175–187.

Keil, A.H., 1973, *Status of Engineering in the Ocean Environment Today*. Cambridge, Mass.: Massachussetts Institute of Technology Sea Grant Rept No. 74-6, 17 pp.

Kukla, G.J., and H.J. Kukla, 1974, Increased surface albedo in the northern hemisphere: *Science*, v. 183, p. 709–714.

Lavi, A., (Ed.). 1973, *Proceedings, Solar Sea Power Plant and Workshop*. Pittsburgh: Pittsburgh, Carnegie-Mellon University, 280 pp.

Lawton, F.L., 1974, Time and tide: *Oceanus*, v. 17, p. 30–37.

Levenspiel, O., and N. De Nevers, 1974, The osmotic pump: *Science*, v. 183, p. 157–160.

McCormick, M.E., R.L. Holt, and C.E. Bosworth, 1975, A pneumatic wave-energy converter for offshore structures: *Offshore Technol. Conf. Proc.*, v. 2, p. 167–170.

Mann, K.H., 1973, Seaweeds: Their productivity and strategy for growth: *Science*, v. 182, p. 975–981.

Martin, M.D., 1974, Power from ocean waves: American Society of Mechanical Engineers, winter annual meeting, New York. Paper 74-WA Pwr-5, 8 pp.

Metz, W.D., 1973, Ocean temperature gradients: Solar power from the sea: *Science*, v. 180, p. 1266–1267.

Metz, W.D., 1977, Ocean thermal energy: The biggest gamble in solar power: *Science*, v. 198, p. 178–180.

Miller, G.R., 1971, Sewage disposal with desalting plant effluent: Undersea Technology, March, p. 17–18.

Newton, J.W., 1976, Photoproduction of molecular hydrogen by a plant–algal symbiotic system: *Science*, v. 191, p. 559–561.

Norman, R.S., 1974, Water salination: A source of energy: *Science*, v. 186, p. 350–352.

Oceanus—1976, Marine Biomedicine, W. Mac Leisch, ed, no 2 Vol. 19. Woods Hole, Mass.: Woods Hole Oceanographic Institution, 60 pp.

*Oswald, W.J., 1974, Solar energy fixation with algal–bacterial systems: *Compost Sci.*, Jan/Feb p. 20–21.

Othmer, D.F., 1969, Desalination of seawater: *In:* Firth, F.E., Ed., *Encyclopedia of Marine Resources*. New York: Van Nostrand Reinhold, p. 162–169.

Othmer, D.F., and O.A. Roels, 1973, Power, fresh water and food from cold, deep sea water: *Science*, v. 182, p. 121–125.

Panicker, N.N., 1976, Power resource estimate of ocean surface waves: *Ocean Eng.*, v. 3, p. 429–439.

Pushkar, N.K., and C.J. Sindermann (Eds.), 1978, *Drugs and Food from the Sea—Myth or Reality?* Univ. Oklahoma Press, 450 pp.

*Richards, A.F., 1976, Extracting energy from the oceans: a review: *Marine Technol. Soc. Jour.*, v. 10, p. 5–21.

Roels, O.A., J.S. Babb, G.L. Mann, and K.C. Haines, 1973, Mariculture in an artificial upwelling system: Paper presented at Fifth Annual Offshore Technology Conference, Houston, Texas, 29 April–2 May, 8 pp.

*Roels, O.A., K.C. Haines, and J.B. Sunderlin, 1975, The potential yield of artificial upwelling mariculture: *10th European Symposium on Marine Biology*, 1975, v. 1, p. 381–390.

*Ruggieri, G.D., 1976, Drugs from th sea: *Science*, v. 194, p. 491–497.

Salter, S.H., 1974, Wave power: *Nature*, v. 249, p. 720–724.

*Scholz, A., T. Horrall, M. Ross, J.C. Cooper, and A. Hasler, 1976, Imprinting two chemical cues: The basis for more stream selection in salmon: *Science*, v. 192, p. 1247–1249.

Science, 1974, Long-Range Weather Forecasting: Sea Temperature Anomalies, vol. 184, pp. 1064–1065.

Sheets, H.E., 1975, Power generation from ocean currents: *Naval Eng. Jour.* v. 87 p. 47–56.

*Silva, A.J., 1977, Physical processes in deep-sea clays: *Oceanus*, v. 20, p. 31–40.

Spilhaus, A., 1972, Bountiful grants of the Sea. Cambridge, Mass.: First Annual Sea Grant Lecture, Massachussets Institute of Technology Sea Grant Rept. No. 73-1, 12 pp.

Steelman, G.E., 1974, An invention designed to convert ocean currents into useable power: *In:* Stewart, H.B. Jr., Ed., *Proceedings of the MacArthur Workshop on the Feasibility of Extracting Useable Energy From the Florida Current.*, Miami: NOAA Atlantic Oceanographic and Meteorological Laboratories, p. 258–277.

Stewart, H.B., 1974, Current from the current: *Oceanus*, v. 17, p. 38–41.

Swift-Hook, D.T., B.M. Count, and I. Glendenning, 1975, Characteristics of a rocking wave power device: *Nature*, v. 254, p. 504–506.

Task Force on Energy, 1974, *U.S. Energy Prospects: An Engineering Viewpoint.* Washington, D.C.: National Academy of Engineering, 141 pp.

Thomas, L., 1976, Marine models in modern medicine: *Oceanus*, v. 19, p. 2–5.

Todd, J.H., J. Atema, and D.B. Boylan, 1972, Chemical communication in the sea: *Marine Technology Society*, v. 6, p. 54–56.

Valent, P.J., and H.J. Lee, 1976, Feasibility of subseafloor emplacement of nuclear waste: *Marine Geotechnol.*, v. 1, p. 267–293.

von Arx, W.S., 1974a, "Energy": Its natural limits and abundances: *In* Stewart, H.B. Jr., Ed., *Proceedings of the MacArthur Workshop on the Feasibility of Extracting Useable Energy from the Florida Current.* Miami: NOAA Atlantic Oceanographic and Meteorological Laboratories, p. 10–21.

von Arx, W.S., 1974b, Energy: Natural limits and abundances: *Oceanus*, v. 17, p. 2–13.

*von Arx, W.S., 1974c, Energy: Natural Limits and Abundances, *EOS*, vol. 55, pp. 828–832.

von Arx, W.S., H.B. Stewart, and J.R. Apel, 1974, The Florida Current as a potential source of useable energy: *In* Stewart, H.B. Jr., Ed., *Proceedings of the MacArthur Workshop on the Feasibility of Extracting Useable Energy from the Florida Current.* Miami: NOAA Atlantic Oceanographic and Meteorological Laboratories, p. 91–103.

*Weinstein, J.N., and F.B. Leitz, 1976, Electric power from differences in salinity: The dialytic battery: *Science*, v. 191, p. 557–559.

Wyrtki, K., E. Stroup, W. Patzert, R. Williams, and W. Quinn, 1976, Predicting and observing El Niño: *Science*, v. 191, p. 343–346.

Zener, C., 1976, Solar sea power: *Bull. Atomic Sci.*, January, p. 17–24.

Zener, C., and J. Fetkovich, 1975, Ocean thermal gradient hydraulic power plant: *Science*, v. 189, p. 293–295.

Index